Communications
in Computer and Information Science   549

Commenced Publication in 2007
Founding and Former Series Editors:
Alfredo Cuzzocrea, Dominik Ślęzak, and Xiaokang Yang

More information about this series at http://www.springer.com/series/7899

Nargozy Danaev · Yurii Shokin
Darkhan Akhmed-Zaki (Eds.)

Mathematical Modeling of Technological Processes

8th International Conference, CITech 2015
Almaty, Kazakhstan, September 24–27, 2015
Proceedings

Editors
Nargozy Danaev
Al-Farabi Kazakh National University
Almaty
Kazakhstan

Yurii Shokin
Institute of Computational Technologies
  SB RAS
Novosibirsk
Russia

Darkhan Akhmed-Zaki
Al-Farabi Kazakh National University
Almaty
Kazakhstan

ISSN 1865-0929   ISSN 1865-0937 (electronic)
Communications in Computer and Information Science
ISBN 978-3-319-25057-1   ISBN 978-3-319-25058-8 (eBook)
DOI 10.1007/978-3-319-25058-8

Library of Congress Control Number: 2015951426

Springer Cham Heidelberg New York Dordrecht London

Springer International Publishing AG Switzerland is part of Springer Science+Business Media
(www.springer.com)

Preface

The International Scientific and Practical Conference "Computational and Information Technologies in Science, Engineering and Education" (CITech) has a long and rich tradition and has been held regularly since 2002.

Historically, the conference was organized in close cooperation between Russian and Kazakh scientists and the general area of discussion was the most advanced achievements in the field of computational technology.

The geographic reach of the conference later expanded and now it is attended by leading scientists from Europe, the USA, Japan, India, and Turkey, among others.

The purpose of the conference is the dissemination of new knowledge and scientific advances among the participants. A special feature of this conference is to involve young scientists in the assessment of their scientific achievements through their interaction with the two countries' leading scientific. Participating in CITech has helped formed a community of new-generation young scientists who are currently conducting important research in the field.

CITech has been held in Almaty (2002, 2004, 2008, 2015), Pavlodar (2006), and Ust-Kamenogorsk (2003, 2013). An important role in the formation of stable traditions for organizing and conducting CITech is played by the personal friendships of scientists from the Novosibirsk Scientific school, such as Prof. Sh. Smagulov, N. Danaev, Yi. Shokin, V. Monakhov, B. Zhumagulov, and many others. Unfortunately, some of them are no longer among us, but we will always remember their contribution to science and education and keep their unforgettable image in our hearts.

For the section "Mathematical Modeling of Technological Processes – CITech-2015" we received 56 articles by authors from seven countries. After all papers were reviewed by at least two international reviewers, the top 20 papers were selected for this volume.

We are grateful to the members of the Program and Organizing Committees, the additional reviewers for their help in preparing this publication, the Ministry of Education and Sciences of the Republic of Kazakhstan, and the Closed Joint Stock Company Intel A/O for support in the organization of conference. We hope the papers of CITech 2015 will be interesting for the readers and of values for the scientific community.

September 2015

Nargozy Danaev
Yurii Shokin
Darkhan Akhmed-Zaki

Organization

Program Committee

Program Committee Co-chairs

Nargozy Danaev	Al-Farabi Kazakh National University, Kazakhstan
Michael Resch	High-Performance Computing Center in Stuttgart, Germany
Yurii Shokin	Institute of Computational Technologies SB RAS, Russia

Program Committee

Ualikhan Abdibekov	H.A. Yassawe International Kazakh-Turkish University, Kazakhstan
Haydar Akca	Abu Dhabi University, UAE
Stanislav Antontsev	Center of Mathematics and Fundamental Applications, University of Lisbon, Portugal
Mirsaid Aripov	National University of Uzbekistan, Uzbekistan
Sergei Bautin	Urals State University of Railway Transport, Russia
Thomas Bonisch	High-Performance Computing Center in Stuttgart, Germany
Igor Bychkov	Institute for System Dynamics and Control Theory SB RAS, Russia
Sergei Cherny	Institute of Computational technologies SB RAS, Russia
Boris Chetverushkin	Keldysh Institute of Applied Mathematics RAS, Russia
Vladimir Danilov	Moscow State Institute of Electronics and Mathematics, Russia
Petkovic Dojcin	University of Pristina, Serbia
Bo Einarsson	Linkoping University, Sweden
Mikhail Fedoruk	Novosibirsk State University, Russia
Anatoly Fedotov	Institute of Computational Technologies SB RAS, Russia
Andreas Griewank	Institut für Mathematik, Humboldt-Universität, Germany
Wagdi George Habashi	McGill University, Canada
Koichi Hayashi	Aoyama Gakuin University, Japan
Valeri Iliyn	Novosibirsk State Technical University, Russia
Simon Jayaraj	National Institute of Technology Calicut, India
Amanbek Jaynakov	Kyrgyz State Technical University named after I. Razzakov, Kyrgyzstan

Ziyaviddin Yuldashev	National University of Uzbekistan (named by after Mirza Ulugbek), Uzbekistan
Yuri Zaharov	Kemerovo State University, Russia
Oleg Zhizhimov	Institute of Computational Technologies SB RAS, Russia
Bakhytzhan Zhumagulov	National Engineering Academy of RK, Kazakhstan

Organizing Committee

Organizing Committee Chairman

| Galimkair Mutanov | Al-Farabi Kazakh National University, Kazakhstan |

Organizing Committee Vice-chairmen

Nargozy Danaev	Al-Farabi Kazakh National Uuniversity, Kazakhstan
Maksat Kalimoldayev	Institute of Information and Computational Technologies, Kazakhstan
Tlekkabul Ramazanov	Al-Farabi Kazakh National Uuniversity, Kazakhstan

Organizing Committee Secretaries

| Denis Esipov | Institute of Computational Technologies SB RAS, Russia |
| Lyazzat Dairbayeva | Al-Farabi Kazakh National University, Kazakhstan |

Organizing Committee Members

Abugamul Abdibekov	Al-Farabi Kazakh National University, Kazakhstan
Darkhan Akhmed-Zaki	Al-Farabi Kazakh National University, Kazakhstan
Timur Bakibayev	Al-Farabi Kazakh National University, Kazakhstan
Maktagali Bektemesov	Al-Farabi Kazakh National University, Kazakhstan
Ernar Imangaliev	Al-Farabi Kazakh National University, Kazakhstan
Baltabek Kanguzhin	Al-Farabi Kazakh National University, Kazakhstan
Almatbek Kidirbekuly	Al-Farabi Kazakh National University, Kazakhstan
Saltanbek Muhambetjanov	Al-Farabi Kazakh National University, Kazakhstan
Ludmila Onishenko	Institute of Information and Computational Technologies, Kazakhstan
Baydaulet Urmashev	Al-Farabi Kazakh National University, Kazakhstan
Irina Vaseva	Institute of Computational Technologies SB RAS, Russia
Dauren Zhakebaev	Al-Farabi Kazakh National University, Kazakhstan
Farkhad Yakhiyayev	Al-Farabi Kazakh National University, Kazakhstan

Additional Reviewers

Haydar Akca	Abu Dhabi University, UAE
Stanislav Antontsev	Center of Mathematics and Fundamental Applications, University of Lisbon, Portugal
Igor Bychkov	Institute for System Dynamics and Control Theory SB RAS, Russia
Sergei Cherny	Institute of Computational Technologies SB RAS, Russia
Nargozy Danaev	Al-Farabi Kazakh National University, Kazakhstan
Vladimir Danilov	Moscow State Institute of Electronics and Mathematics, Russia
Mikhail Fedoruk	Novosibirsk State University, Russia
Andreas Griewank	Institut für Mathematik, Humboldt-Universität, Germany
Koichi Hayashi	Aoyama Gakuin University, Japan
Valeri Iliyn	Novosibirsk State Technical University, Russia
Simon Jayaraj	National Institute of Technology Calicut, India
Christophe Josserand	Institute Jean Le Rond D'Alembert, Paris
Sergey Kabanikhin	Institute of Computational Mathematics and Mathematical Geophysics SB RAS, Russia
Aidarkhan Kaltayev	Al-Farabi Kazakh National University, Kazakhstan
Shoshana Kamin	University of Tel Aviv, Israel
Robert Kersner	University of Pecs, Hungry
Stanislav Kharin	Kazakh-British Technical University, Kazakhstan
Matthias Meinke	Institute of Aerodynamics, RWTH University, Germany
Anvarbek Meirmanov	Kazak-British Technical University, Kazakhstan
Wolfgan Merkle	Heidelberg University, Institute of Mathematics and Computer Science, Germany
Givi Peyman	University of Pittsburgh, USA
Alexander Prokopenya	Warsaw University of Life Sciences, Poland
Michael Resch	High-Performance Computing Center in Stuttgart, Germany
Boris Ryabko	Siberian State University of Telecommunications and Information Sciences, Russia
Vladimir Shaidurov	Institute of computational modelling SB RAS, Russia
Yurii Shokin	Institute of computational technologies SB RAS, Russia
Sergei Turitsyn	Aston University, UK
Lian-Ping Wang	University of Delaware, USA
Bakhytzhan Zhumagulov	National Engineering Academy of RK, Kazakhstan

Contents

Mathematical Modelling of Oil Recovery by Polymer/Surfactant Flooding

Nargozy Danaev, Darkhan Akhmed-Zaki$^{(\boxtimes)}$, Saltanbek Mukhambetzhanov, and Timur Imankulov

Al-Farabi Kazakh National University, Al-Farabi ave., 71, Almaty, Kazakhstan
{darhan_a,mukhambetzhanov_,imankulov_ts}@mail.ru

Abstract. This article describes a hydrodynamic model of collaborative fluids (oil, water) flow in porous media for enhanced oil recovery, which takes into account the influence of temperature, polymer and surfactant concentration changes on water and oil viscosity. For the mathematical description of oil displacement process by polymer and surfactant injection in a porous medium, we used the balance equations for the oil and water phase, the transport equation of the polymer/surfactant/salt and heat transfer equation. Also, consider the change of permeability for an aqueous phase, depending on the polymer adsorption and residual resistance factor. Results of the numerical investigation on three-dimensional domain are presented in this article and distributions of pressure, saturation, concentrations of polymer/surfactant/salt and temperature are determined. The results of polymer/surfactant flooding are verified by comparing with the results obtained from ECLIPSE 100 (Black Oil). The aim of this work is to study the mathematical model of non-isothermal oil displacement by polymer/surfactant flooding, and to show the efficiency of the combined method for oil-recovery.

Keywords: EOR · Polymer · Surfactant · Darcy · Porous media · MPI

1 Introduction

The investigations show that the use of chemical methods for increasing oil recovery, such as polymer and surfactant flooding are the effective chemical EOR methods. There are various interactions between the surfactant and the reservoir fluids, such as adsorption, interfacial tension, wettability [1]. Surfactants are used to reduce the interfacial tension between crude oil and reservoir water and increase the mobility of "trapped" oil in the pore space. Polymer injection method used for enhancing the efficiency of displacement by reducing mobility and increasing viscosity of water phase [2,3]. At present, the combined methods of enhancing oil recovery are used. One of such methods is surfactant flooding in combination with water soluble polymers. Surfactant and polymer are injected into the reservoir, then displace oil to the production wells by pumping water. When using this method, the oil recovery rate is higher in comparison with the

© Springer International Publishing Switzerland 2015
N. Danaev et al. (Eds.): CITech 2015, CCIS 549, pp. 1–12, 2015.
DOI: 10.1007/978-3-319-25058-8_1

method when surfactant and polymer are used separately [4,5]. The aims of this work: 1) to study the mathematical model of oil displacement by polymer-surfactant flooding, which is considers the influence of temperature effects and dependence of polymer/surfactant solution viscosity on agents concentration and water salinity; 2) to develop a sequential/parallel computational algorithm for solution of 3D problem using MPI technology; 3) study of oil recovery factor at different impact on the reservoir.

2   Mathematical Model

In a general case, displacement of oil by polymer and surfactant is effected by complex physico-chemical processes, when modeling and numerical realization of which there take place definite problems. For example, viscosity of injected solution depends on various factors, such as reservoir temperature, concentration of polymer/surfactant in solution and water salinity and etc. The model takes into account the following assumptions:

- the porous media and fluid are incompressible;
- gravitational forces are not taken into account;
- the two-phase flow (aqueous, oleic) is subject of the Darcyś law;
- water, polymer, surfactant and salt are fully mixed;
- adsorption of the polymer affects only on the relative permeability of the aqueous phase;
- dissolution of polymer and salt in oil is very small.

Based on the above mentioned assumptions, we can write the mathematical model of two-phase flow in porous media. Mass conservation equation for aqueous and oleic phases [6] is:

$$m\frac{\partial s_w}{\partial t} + div(\boldsymbol{v}_w) = q_1 \qquad (1)$$

$$m\frac{\partial s_o}{\partial t} + div(\boldsymbol{v}_o) = q_2 \qquad (2)$$

$$s_w + s_o = 1$$

where m - porosity, s_w, s_o - water and oil saturations, q_1, q_2 - source or sink, $\boldsymbol{v}_w, \boldsymbol{v}_o$ - velocities of the water and oil phases which is expressed by the following law:

$$\boldsymbol{v}_i = -K_0 \frac{f_i(s)}{\mu_i} \bigtriangledown P, \quad i = w, o \qquad (3)$$

$f_i(s), \mu_i$ - relative permeability and viscosity of fluids, K_0 - absolute permeability.

Polymer, surfactant and salt transport equations can be written as [1]:

$$m\frac{\partial}{\partial t}(c_p s_w) + \frac{\partial a_p}{\partial t} + div(\boldsymbol{v}_w c_p) = div(m D_{pw} s_w \bigtriangledown c_p) \qquad (4)$$

$$m\frac{\partial}{\partial t}(c_{sw} s_w + c_{so} s_o) + \frac{\partial a_{surf}}{\partial t} + div(\boldsymbol{v}_w c_{sw}) + div(\boldsymbol{v}_o c_{so}) =$$

$$= div(mD_{sw}s_w \nabla c_{sw} + mD_{so}s_o \nabla c_{so}) \tag{5}$$

$$m\frac{\partial}{\partial t}(c_s s_w) + div(\boldsymbol{v}_w c_s) = 0 \tag{6}$$

where c_p, c_s - polymer and salt concentrations in aqueous phase, c_{sw}, c_{so}- surfactant concentration in aqueous and oleic phases, a_p, a_{surf} - polymer and surfactant adsorption functions, D_{pw}, D_{sw}, D_{so} - polymer and surfactant diffusion coefficients.

Heat transfer equation:

$$\frac{\partial}{\partial t}(((1-m)C_r\rho_r + m(C_w s_w \rho_w + C_o s_o \rho_o))T) + div(\rho_w C_w \boldsymbol{v}_w T) + div(\rho_o C_o \boldsymbol{v}_o T) =$$

$$= div((1-m)\lambda_0 + m(\lambda_1 s_w + \lambda_2 s_0) \nabla T) \tag{7}$$

where C_w, C_o, C_r - specific heat of water, oil and rock, ρ_w, ρ_o, ρ_r - density of water, oil and rock, $\lambda_w, \lambda_o, \lambda_r$ - coefficients of thermal conductivity.

Flory-Huggins equation can represent a mathematical relation, which describes the dependence of water phase viscosity on the concentration of salt, surfactant and polymer. This dependence which takes into account temperature changes can be written as [7]:

$$\mu_a = \mu_w(1 + (\gamma_1 c_p + \gamma_2 c_p^2 + \gamma_3 c_{sw} + \gamma_4 c_{sw}^2)c_s^{\gamma_5} - \gamma_6(T - T_p)) \tag{8}$$

$$\mu_o = \mu_{0_0}(1 - \gamma_7(T - T_p)) \tag{9}$$

where $\gamma_1, \gamma_2, \gamma_3, \gamma_4, \gamma_5, \gamma_6, \gamma_7$ - nondimensional constants, μ_{0_0} - initial viscosity of oelic phase, T_p - reservoir temperature.

Relative permeability curves are taken as follows:

$$f_w(s_w) = s_w^{3.5}; \quad f_o(s_w) = (1 - s_w)^{3.5}$$

The type of the polymer and surfactant determines their adsorptions degree. Langmuirs law can represent the relation between adsorbed polymer/surfactant concentration in the solution [1]:

$$a_p = \frac{b_1 c_p}{1 + b_1 c_p}, \quad a_{surf} = \frac{b_2 c_{sw}}{1 + b_2 c_{sw}}$$

where b_1, b_2 - Langmuirś constants.

Permeability reduction factor R_k can be described as follows [8]:

$$R_k = 1 + (R_{RF} - 1)a_p$$

R_{RF}- residual reduction factor.

Initial and boundary conditions are:

$$s_w|_{t=0} = s_{w0}, \quad c_{pw}|_{t=0} = c_{p0}, \quad a_p|_{t=0} = a_{p0}$$

$$c_{sw}|_{t=0} = c_{sw0}, \quad c_{so}|_{t=0} = c_{so0}, \quad a_{surf0}|_{t=0} = a_{surf0} \tag{10}$$

$$c_s|_{t=0} = s_{s0}, \quad T|_{t=0} = T_p$$

$$\frac{\partial s_w}{\partial n}|_{\partial\Omega} = 0; \quad \frac{\partial P}{\partial n}|_{\partial\Omega} = \gamma V_p; \quad \frac{\partial T}{\partial n}|_{\partial\Omega} = \gamma V_c;$$

$$-D\frac{\partial c_{pw}}{\partial n} + \boldsymbol{v}_{1n} c_{pw}|_{\partial\Omega} = q_n \tilde{c_{pw}}; \tag{11}$$

$$-D\frac{\partial c_{sw}}{\partial n} + \boldsymbol{v}_{1n} c_{sw}|_{\partial\Omega} = q_n \tilde{c_{sw}}; \quad -D\frac{\partial c_s}{\partial n} + \boldsymbol{v}_{1n} c_s|_{\partial\Omega} = 0;$$

Pressure equation obtained by adding (1) and (2):

$$div(\boldsymbol{v}_w) + div(\boldsymbol{v}_o) = q_1 + q_2 \tag{12}$$

3   Numerical Method

For numerical calculation, consistency of units and order of variables are important. Therefore, a system of equations (1) - (12) is converted to a dimensionless form. To solve these equations, an explicit/implicit scheme is used [10]. First of all, fluid properties and physical parameters of reservoir are set. Further calculations are conducted in the following order:

- distribution of pressure (capillary pressure);
- saturation (by the known distribution of pressure);
- distribution of salt, surfactant and polymer concentrations;
- distribution of temperature in the reservoir;
- aqueous phase viscosity, depending on salt, surfactant and polymer concentrations is recalculated;
- aqueous phase permeability considering the polymer adsorption is recalculated.

Table 1 gives a information about influence of polymer and surfactant concentrations on the main parameters of mathematical model of oil displacement process by polymer and surfactant solutions. The table shows that both polymer and surfactant effect viscosities of the both phases and do not effect the relative permeabilities. Capillary pressure takes into account the influence of surfactant concentration and the absolute permeability of rock decreases during injection of the polymer.

Table 1. Influence of polymer and surfactant concentrations on the main parameters.

	Polymer	Surfactant
Capillary effects, P_c	−	+
Relative pearmeabilities, f_w, f_o	−	−
Phase viscosities μ_a, μ_b	+	+
Absolute permeability K_0	+	−
Adsorption, a	+	+

4   Computational Results

The results of numerical calculations for non-isothermal oil displacement are shown in Figures 1-6.

Figure 1a shows the permeability distribution. It can be noted, that for calculation of distribution of the main parameters used heterogeneous field. In opposite corners of the selected area are two wells: injection and production. These wells are set bottom hole pressure (P_{inj} or P_{prod}). Figure 1b shows the results of calculating the distribution of pressure in domain. Distribution of water saturation, polymer and surfactant concentrations, which are pumped through injection well, presented in Figures 2, 3 and 4. It is considered that the salinity of injection water is equal to zero (Figure 5). In these calculations, the solution is pumped into the reservoir over the reservoir temperature, the distribution of which is shown in Figure 6. Thus, the problem is solved numerically in a simple formulation, i.e. not taken into account changes in viscosity of the concentration of the reagents and temperature, polymer adsorption was not affected by the

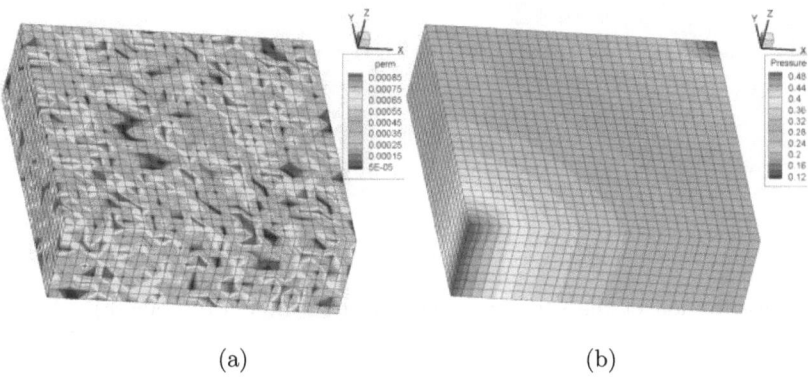

(a)   (b)

Fig. 1. (a) Permeability field; (b) distribution of pressure.

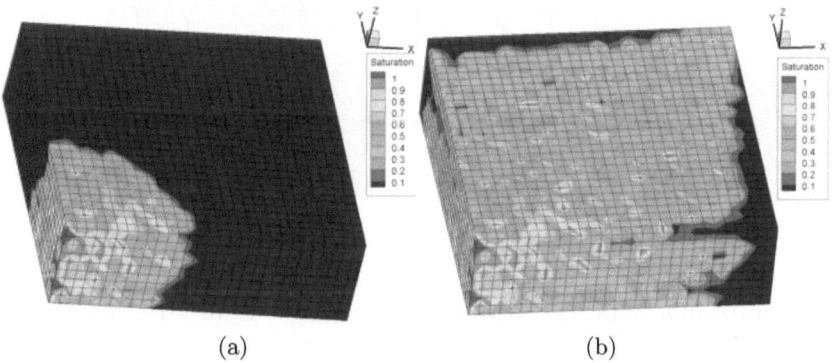

(a)   (b)

Fig. 2. Distribution of water saturation after: (a) - 50; (b) - 150 time iterations

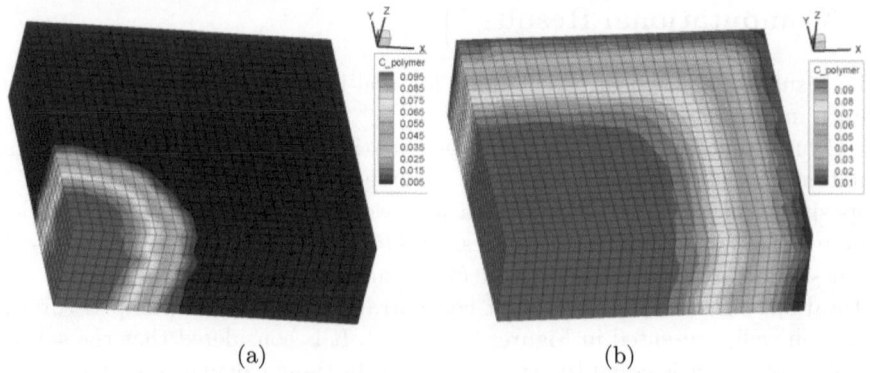

Fig. 3. Distribution of the polymer concentration after: (a) - 50; (b) - 150 time iterations

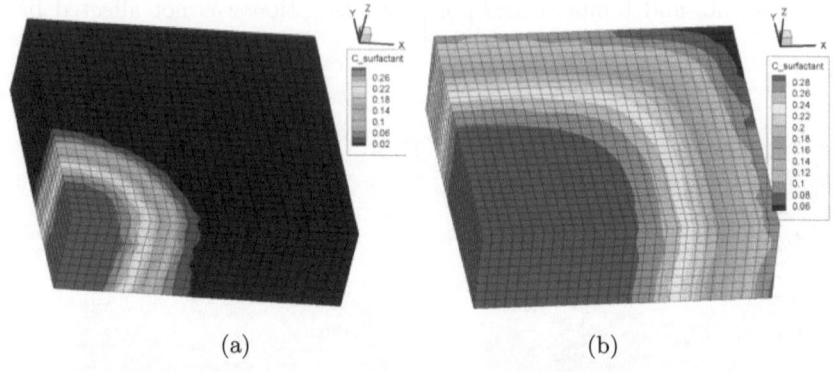

Fig. 4. Distribution of the surfactant concentration after: (a) - 50; (b) - 150 time iterations

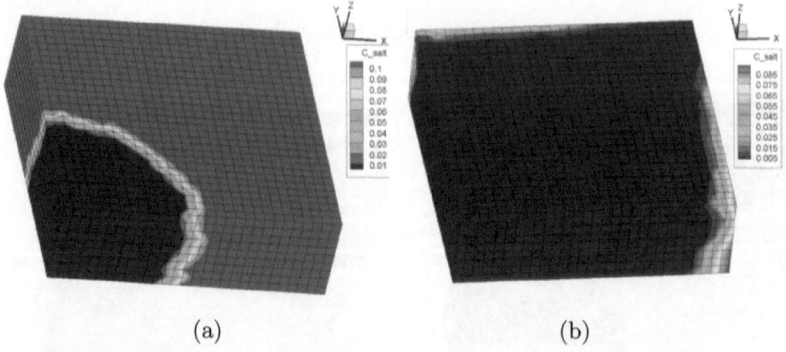

Fig. 5. Distribution of the salt concentration after: (a) - 50; (b) - 150 time iterations

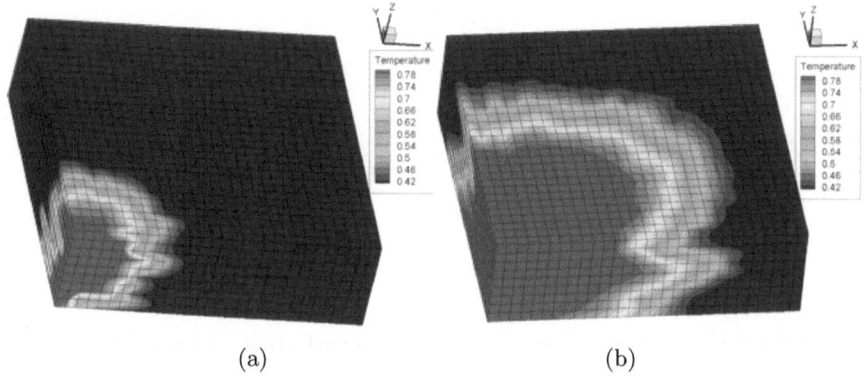

Fig. 6. The temperature distribution after: (a) - 50; (b) - 150 time iterations

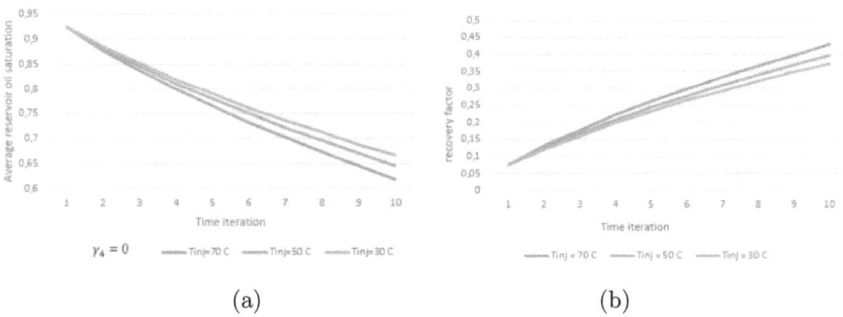

Fig. 7. Average reservoir oil saturation (a), recovery factor (b).

permeability of the aqueous phase. In further calculations to get technological parameters they are taken into account.

Figure 7 shows the influence of injected water temperature on average oil saturation in reservoir and recovery factor. It may be noted that injection of polymer solution at temperatures above the reservoir indicates a higher displacement efficiency at a certain time.

Figure 8a shows the variation of the oil recovery at different impact on the reservoir: oil displacement by water, displacement by using a surfactant and oil displacement by polymer solution. Naturally, use of chemical reagents shows higher oil recovery than using water. It may be noted that the surfactant solution to a certain point of time shows a high recovery factor, but after about 90 time iterations it is relatively worse. The polymer solution shows a high recovery factor for the whole period of operation. From an economic point of view, we can not always inject these reagents. For this reason, used the following sequence of chemical injection into the reservoir (see Figure 9): at first surfactant solution is injected, which displace the oil and reduces the interfacial tension between "trapped" oil and water, because of capillary pressure. Then all this displaced

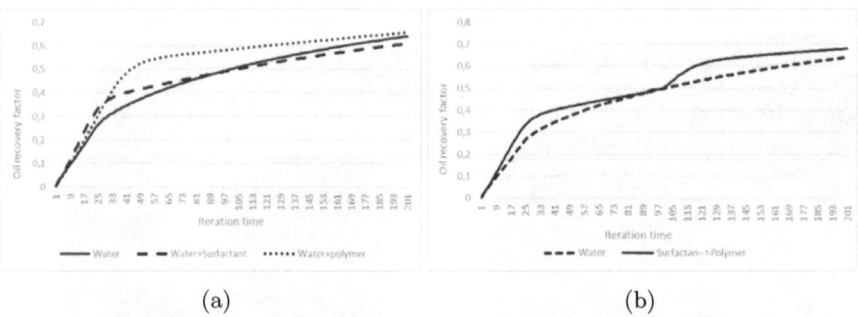

(a)	(b)

Fig. 8. Oil recovery factor: (a) water, polymer and surfactant solutions; (b) combined flooding (polymer+surfactant).

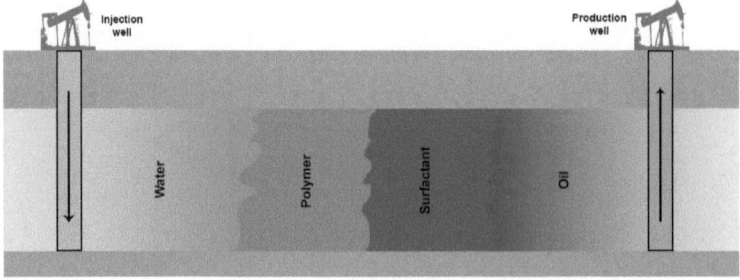

Fig. 9. Sequential injection of chemicals.

by more viscous polymer solution. It is known that polymer flooding increases oil recovery by increasing water viscosity. Then all this displaced by water. When using this sequence of injection, it is important to know when to stop adding surfactant and start polymer injection to obtain a higher recovery factor. In these calculations, after about 70 time iterations begins polymer injection (respectively stops injection of surfactant). After some time oil recovery rising again because the surface tension between the phases has fallen and all this displaced by relatively more viscous solution, which can be seen in Figure 8b. Of course, it would be good to calculate the optimal and cost-effective concentrations of injected agents to achieve the maximum oil recovery. But with these studies we can say that the use of hybrid technology of chemical flooding (in this case the surfactant + polymer) yields positive result.

Calculation of this model on the grid 64x64x64 and more takes a huge amount of time. Therefore, it would be advisable to use of parallelization technology to achieve high performance computation. For parallelization of this algorithm, the computational domain is divided into partially overlapping subdomains, calculations in which are performed independently of each other. After each iteration, it is necessary to make the exchange of data at the boundaries of the subdomains [11]. The above method was implemented using MPI tech-

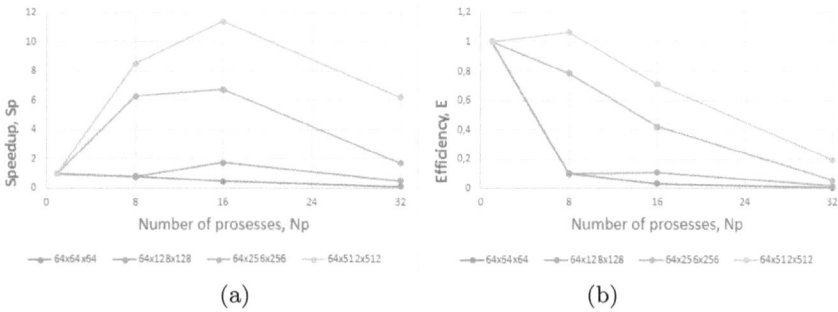

Fig. 10. Comparison of speedup (a) and efficiency (b) of parallel algorithm for different grid sizes.

nology. The speedup and efficiency of parallel algorithm on the 64x64x64, 64x128x128, 64x256x256, 64x512x512 grids are investigated. The results are shown in Figure 10. These graphs show that high efficiency can only be achieved on large grids.

5 Model Verification

Correctness of proposed model was confirmed by two stages of verification [9]:

- comparison of numerical results with laboratory experiments;
- and with results of calculations on hydrodynamic simulator Eclipse 100.

First stage. Verification of mentioned above model is based on the results of a laboratory experiment conducted by research group of Engineering specialization Laboratory, leaded by Kudaibergenov S.E. Investigation of oil displacement in cores with water and polymer solution performed on UIC-C(2) [12] installation. Input data for numerical simulation of this process (which is fully consistent with experimental data) are shown in Table 2.

Figure 11 shows dependence of oil displacement on injected pore volume, obtained by numerical and laboratory research. It can be noted that, oil displacement by polymer shows much higher recovery ratio compared with water

Table 2. Physical parameters used in simulation.

Parameter	Value
Porosity, m	0.37
Absolute permeability K_0	0.322 Darcy
RRF, R_{RF}	1.2
Concentration of injection Gellan Solution C_{inj}	0.1 %
Oil Viscosity μ_o	8.09 mPa $\cdot$ s
Water Viscosity μ_w	0.9 mPa $\cdot$ s
Adsorption Constant	0.1 m^3/kg
Initial Salt Concentration (NaCl) C_{init_salt}	73 g/l

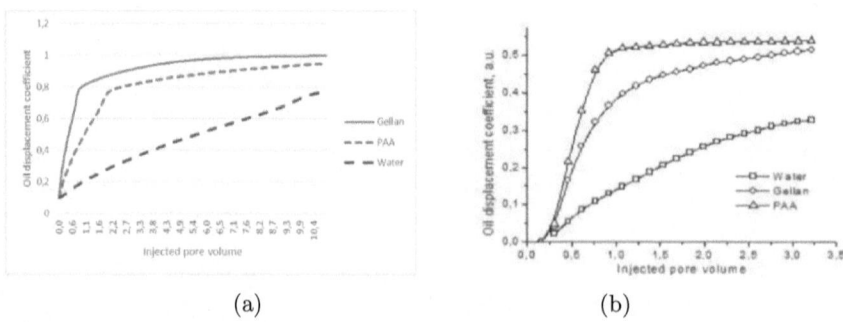

(a)   (b)

Fig. 11. Dependence of recovery rate on injected pore volume of fluid. a) numerical, b) experimental study.

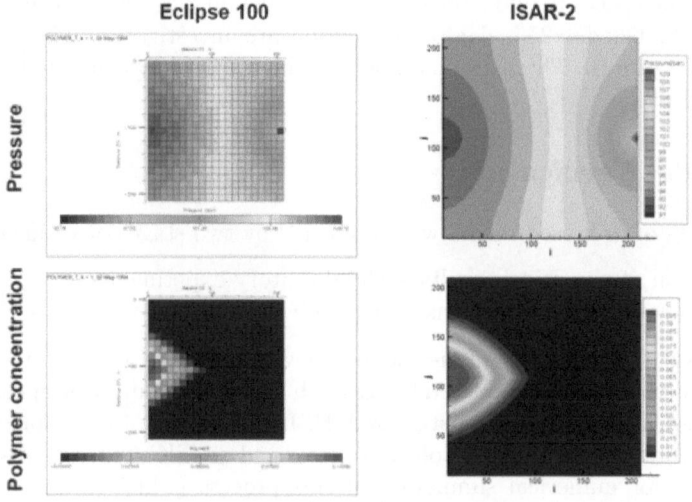

Fig. 12. Distributions of pressure and polymer concentration before water breakthrough.

displacement. Oil displacement efficiency of gellan and polyacrylamide at about the same level, which confirming results of the experimental study.

Second stage. To compare numerical results with simulator Eclipse 100 two dimensional problem is considered. Table 3 shows fluids properties and reservoir parameters for numerical modeling. It is exactly the same values used in the calculation on the Eclipse simulator. Figure 12 and 13 shows simulation results of main parameters using Eclipse 100 (first column) and proposed model (second column).

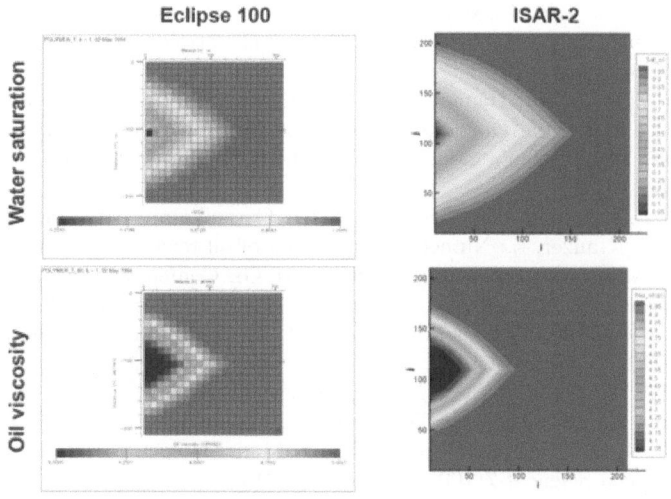

Fig. 13. Distributions of water saturation and oil viscosity before water breakthrough.

6   Conclusion

A mathematical model of oil displacement process by polymer-surfactant injection is considered and solved, taking into account the dependence of solution viscosity on salt, surfactant and polymer concentrations and where viscosity of both phase are depends on temperature. System of equations is solved using implicit/explicit methods and following numerical results were obtained: distribution of pressure, saturation distribution of both phases, salt, surfactant and polymer concentration and temperature distribution in reservoir. A serial / parallel computing algorithm for solving the three-dimensional problem is developed and the efficiency and speedup of algorithm for different grid sizes compared. The efficiency of oil displacement by using a combination of flooding with polymer and surfactant is showed. The polymer and surfactant injection process into the oil reservoir for enhanced oil recovery can be modeled using the proposed model. The presented results show good consistence compared with the results of the hydrodynamic simulator Eclipse (Black Oil).

References

1. Babalyan, G.A., Levy, B.I., Tumasyan, A.B., Khalimov, E.M.: Oilfield development using surfactants. Nedra, Moscow (1983)
2. Lake, L.W.: Enhanced oil recovery. Prentice Hall Inc, Englewood Cliffs (1989)
3. Sorbie, K.S.: Polymer improved oil recovery. CRC Press, Boca Raton (1991)
4. Fathaddin, M.T., Nugrahanti, A., Buang, N.P., Elraes, K.A.: Surfactant-polymer flooding performance in heterogeneous two-layered porous media. IIUM Engineering Journal **12**(1), 31–38 (2011)

5. Rai, K., Johns, T.R., Delshad, M., Lake, W.L., Goudarzi, A.: Oil-recovery predictions for surfactant polymer flooding. Journal of Petroleum Science and Engineering. **112**, 341–350 (2013)
6. Akhmed-Zaki, D.Zh., Danaev, N.T., Mukhambetzhanov, S.T., Imankulov, T.S.: Analysis and evaluation of heat and mass transfer processes in porous media based on Darcy-Stefan's model. In: ECMOR XIII, Biarritz, France, p. 122, 10–13 September 2012
7. Flory, P.J.: Principles of polymer chemistry. Cornell University Press (1953)
8. Wegner, J., Ganzer, L.: Numerical simulation of oil recovery by polymer injection using COMSOL. In: Proceeding of the COMSOL Conference, Milan (2012)
9. Imankulov, T.S., Lebedev, D., Aidarov, K., Turar, O.: Design of HPC system for analysis the gel-polymer flooding of oil fields. Contemporary Engineering Sciences **7**(27), 1531–1545 (2014)
10. Samarskii, A.A., Gulin, A.V.: Numerical methods. Nedra, Moscow (1989)
11. Matkerim, B., Akhmed-Zaki, D., Barata, M.: Development of High Performance Scientific Computing Application Using Model-Driven Architecture. Applied Mathematical Sciences **7**(100), 4961–4974 (2013)
12. Ibragimov, R., Gusenov, I., Tatykhanova, G., Adilov, Zh, Nuraje, N., Kudaibergenov, S.: Study of Gellan for Polymer Flooding. Journal of Dispersion Science and Technology **9**, 1240–1247 (2013)

Modelling of Evolution Small-Scale Magnetohydrodynamic Turbulence Depending on the Magnetic Viscosity of the Environment

Aigerim Abdibekova[1](✉), Bakhytzhan Zhumagulov[2], and Dauren Zhakebayev[1](✉)

[1] Al-Farabi Kazakh National University, Al-Farabi ave. 71, 050040 Almaty, Kazakhstan
a_aigerim@inbox.ru, daurjaz@mail.ru
[2] National Academy of Engineering of the Republic Kazakhstan, Bogenbai Batyr str. 80, 050010 Almaty, Kazakhstan

Abstract. The present work is devoted to study of self-excitation of magnetic field and the motion of the conducting fluid at the same time taking into account acting forces. The idea is to specify in the phase space of initial conditions for the velocity field and magnetic field, which satisfy the condition of continuity. Given initial condition with the phase space is translated into physical space using a Fourier transform. The obtained velocity field and magnetic field are used as initial conditions for the filtered MHD equations. Further is solved the unsteady three-dimensional equation of magnetohydrodynamics to simulate homogeneous MHD turbulence decay.

Keywords: MHD turbulence · Turbulence · Small-scale · LES

1 Introduction

An examination of the homogeneous magnetohydrodynamic turbulence decay process, in spite of the large number of publications in this field, is a relevant task for researchers of several generations. The influence of magnetic field on the conducting fluid is studied in various fields of science and used in an engineering and technology. Therefore, studies of magnetohydrodynamic turbulence decay is an important task in the fields of: forming astrophysical and geophysical phenomena, MHD generators, plasma accelerators and engines. The study of the magnetohydrodynamic (MHD) turbulence process in a small range of change of the Reynolds (Re_m) magnetic number can be modeled and experimentally investigated, while the same process remains beyond experimental reach and computational techniques for a broad range of values. Research problems of the magnetic field depending on the electro conductive fluid is divided into three types:

© Springer International Publishing Switzerland 2015
N. Danaev et al. (Eds.): CITech 2015, CCIS 549, pp. 13–25, 2015.
DOI: 10.1007/978-3-319-25058-8_2

1. An examination of the MHD turbulence at a constant value of the magnetic field.

2. An examination of the self-excitation of magnetic field at a given velocity of the flow.

3. An examination of the self-excitation of magnetic field and the motion of a conducting fluid at the same time taking into account acting forces.

The problem of the magnetic field influence on turbulent flows was first raised by [1], who provided basic equations and an analytical solution for the movement of an electrically conducting fluid. The first numerical study of magnetohydro-dynamic turbulence problem of the first type conducted by [2] at the magnetic number $Re_m \ll 1$. The numerical experiment of Schumann was the reflection of the idea of [3], who researched a homogeneous isotropic flow influenced by an applied external magnetic field. The modeling outlined in the publications of these scientists is performed using a spectral method, which is used as the basis for presenting a quantitative description of magnetic damping, the emergence of anisotropy, and the dependency of the results on the presence or the absence of a non-linear summand in the Navier-Stokes equation. The low performance of computing machines at that time did not permit the full solution of this problem. Later, a similar problem was researched first by [4] and later by [5]. These authors presented the results of direct numerical modeling of large-scale structures in a periodic magnetic field, which reflected a change in the turbu-lence statistical parameters as a result of an imposed magnetic field influence. The contribution of these scientists in this area of expertise is determined by proving that the behavior of two- and three-dimensional structures varies sub-stantially. A similar result was obtained by [6] in examining locally isotropic structures by the method of large eddies. Although the result obtained for the anisotropy invariant distribution and the Reynolds strength was discussed by several researchers, the findings on this matter cannot be considered conclusive because the force of the magnetic field is the determining factor for the change of quantitative indicators of invariants, which was not demonstrated by the author.

A generalization of a linear case researched by [2] and [3] is featured in pub-lications by [7]. These researchers demonstrated a redistribution of the kinetic energy between velocity components, which indicated an inconsistency with a previously presented linear theory. In a nonlinear case, velocity components that are parallel and perpendicular to the magnetic field decay at various velocities, which is an apparent inconsistency with the earlier numerical experiments.

The process of the magnetic field influence on a developed turbulence was examined by [8],and demonstrated the possibility of using the quasi-stationary approximation for the solution of the second type problem and suggested to use quasi-linear approximations to solve the problem at $Re_m = 20$. One of the second type problem results were reported in [9], the modeling of a diminishing MHD turbulence by LES and DNS methods and demonstrated that the magnetic field at the initial time started to decay under the influence of the total kinetic energy. This effect is consistent with Joule dissipation. A similar picture of the decay was not reported by the authors because their main objective was the evaluation the model adequacy for the LES and DNS methods. Accordingly, there was a justification of the modified dynamic Smagorinsky model for simulation of the temporal decaying magnetohydrodynamic turbulence.

The results of the third type of problem was presented by [10], and produced a detailed investigation of pseudospectral direct numerical simulation (DNS), with up to 1024^3 nodes, three-dimensional incompressible magnetohydrodynamic (MHD) turbulence, without the mean magnetic field. Study was carried out according to various statistical properties of the both decreasing and statistically steady MHD turbulence on the magnetic Prandtl number Prm, taken over in a wide range, $0.01 \leq Pr_m \leq 10$. Turbulent characteristics were obtained at a constant magnetic viscosity for different values of the kinetic viscosity.

This work is devoted to study of self-excitation of magnetic field and the motion of the conducting fluid at the same time taking into account acting forces. The idea is to specify in the phase space of initial conditions for the velocity field and magnetic field, which satisfy the condition of continuity. Given initial condition with the phase space is translated into physical space using a Fourier transform. The obtained of velocity field and magnetic field are used as initial conditions for the filtered MHD equations. Further is solved the unsteady three-dimensional equation of magnetohydrodynamics to simulate homogeneous MHD turbulence decay.

2   Problem

The numerical modeling of a homogeneous MHD turbulence decay based on the large eddy simulation method depending on the conductive properties of the incompressible fluid is reviewed.

The numerical modeling of the problem is performed based on solving non-stationary filtered magnetic hydrodynamics equations in conjunction with the continuity equation in the Cartesian coordinate system in a non-dimensional form:

$$
\begin{cases}
\dfrac{\partial(\bar{u}_i)}{\partial t} + \dfrac{\partial(\bar{u}_i\bar{u}_j)}{\partial x_j} = -\dfrac{\partial(\bar{p})}{\partial x_i} + \dfrac{1}{Re}\dfrac{\partial}{\partial x_j}\left(\dfrac{\partial(\bar{u}_i)}{\partial x_j}\right) - \dfrac{\partial(\tau^u_{ij})}{\partial x_j} + A\dfrac{\partial}{\partial x_j}\left(\bar{H}_i\bar{H}_j\right), \\[3ex]
\dfrac{\partial(\bar{u}_j)}{\partial x_j} = 0, \\[3ex]
\dfrac{\partial(\bar{H}_i)}{\partial t} + \dfrac{\partial(\bar{u}_j\bar{H}_i)}{\partial x_j} - \dfrac{\partial(\bar{H}_j\bar{u}_i)}{\partial x_j} = \dfrac{1}{Re_m}\dfrac{\partial}{\partial x_j}\left(\dfrac{\partial(\bar{H}_i)}{\partial x_j}\right) - \dfrac{\partial(\tau^H_{ij})}{\partial x_j}, \\[3ex]
\dfrac{\partial(\bar{H}_j)}{\partial x_j} = 0, \\[3ex]
\tau^u_{ij} = \left((\overline{u_i u_j}) - (\bar{u}_i\bar{u}_j)\right) - \left((\overline{H_i H_j}) - (\bar{H}_i\bar{H}_j)\right), \\[3ex]
\tau^H_{ij} = \left((\overline{u_i H_j}) - (\bar{u}_i\bar{H}_j)\right) - \left((\overline{H_i u_j}) - (\bar{H}_i\bar{u}_j)\right),
\end{cases}
\tag{1}
$$

where $\bar{u}_i$ $(i = 1, 2, 3)$ are the velocity components, $\bar{H}_1$, $\bar{H}_2$, $\bar{H}_3$ are the magnetic field strength components, $A = H^2/(4\pi\rho V^2) = \Pi/Re_m^2$ is the Alfvén number, H is the characteristic value of the magnetic field strength, V is the typical velocity, $\Pi = (V_A L/\nu_m)^2$ is a dimensionless value (on which the value Π depends in the equation for $\bar{H}_i$). If $\Pi << 1$, then $\partial\bar{H}_i/\partial t = 0$. The publication by [11] discussed in detail the physics of phenomena related to the ability to disregard the summand $\partial\bar{H}_i/\partial t$. $(V_A)^2 = H^2/4\pi\rho$ is the Alfvén velocity, $\bar{p} = p + \bar{H}^2 A/2$ is the full pressure, t is the time, $Re = LV/\nu$ is the Reynolds number, $Re_m = VL/\nu_m$ is the magnetic Reynolds number, L is the typical length, ν is the kinematic viscosity coefficient, ν_m is the magnetic viscosity coefficient, ρ is the density of electrically conducting incompressible fluid, and τ_{ij}^u, τ_{ij}^H is the subgrid-scale tensors responsible for small-scale structures to be modeled. To model a subgrid-scale tensor, a viscosity model is presented as $\tau_{ij}^u = -2\nu_T\bar{S}_{ij}$, where $\nu_T = C_S\Delta^2 \left(2\bar{S}_{ij}\bar{S}_{ij}\right)^{\frac{1}{2}}$ is the turbulent viscosity, $\bar{S}_{ij} = (\partial\bar{u}_i/\partial x_j + \partial\bar{u}_j/\partial x_i)/2$ is the deformation velocity tensor value. To model a magnetic subgrid-scale tensor, a viscosity model is used: $\tau_{ij}^H = -2\eta_t\bar{J}_{ij}$, where $\eta_t = D_S\Delta^2 \left(\bar{J}_{ij}\bar{J}_{ij}\right)^{\frac{1}{2}}$ is the turbulent magnetic diffusion, the coefficients C_S, D_S are calculated for each determined time layer, and $\bar{J}_{ij} = (\partial\bar{H}_i/\partial x_j - \partial\bar{H}_j/\partial x_i)/2$ is the magnetic rotation tensor.

Periodic boundary conditions are selected at all borders of the reviewed area of the velocity components and the magnetic field strength.

The initial values for each velocity component and strength are defined in the form of a function that depends on the wave numbers in the phase space:

$$u_i\left(k_i, 0\right) = k_i^{\frac{b-2}{2}} e^{-\frac{b}{4}\left(\frac{k_i}{k_{\max}}\right)^2}; H_i\left(k_i, 0\right) = k_i^{\frac{b-2}{2}} e^{-\frac{b}{4}\left(\frac{k_i}{k_{\max}}\right)^2},$$

where $\bar{u}_i$ is the one-dimensional velocity spectrum, $i = 1$ refers to the longitudinal spectrum, $i = 2$ and $i = 3$ refer to the transverse spectrum, $\bar{H}_i$ is the one-dimensional magnetic field strength spectrum, m is the spectrum power, and k_1, k_2, k_3 are the wave numbers.

For this problem we selected a variational parameter b and the wave number k_{max}, which determine the type of turbulence. For modeling homogeneous MHD turbulence can be set parameters k_{max} and b, which correspond to the experimental data [12].

3   Method for Calculating the Small-Scale Turbulence Coefficient

Along with the accepted calculated grid, a grid with twice the size of cells along each axis is used. The large grid number cell is indicated as p, g, r (p, g, r are the axes numbered x_1, x_2, x_3, respectively), $p = 1, 2, 3, ..., N_1/2$, $g = 1, 2, 3, ..., N_2/2$, and $r = 1, 2, 3, ..., N_3/2$. The cell with the number α along axis x_1 includes the cells of the initial grid with numbers $n = 2p - 1$ and $n = 2p$, where n changes within the range from 1 to N_1. Similar to number g, for x_2 determined cells with numbers $m = 2g - 1$ and $m = 2g$, $q = 2r - 1$ and $q = 2r$. Therefore, one cell p, g, r of a large grid is the same as eight cells of the initial grid.

The average values u_1^2, u_2^2, u_3^2 for the total volume of the calculated area of the liquid flow are marked $\langle u_1 \rangle^2, \langle u_2 \rangle^2, \langle u_3 \rangle^2$. These values can be calculated using smaller and larger calculation grids:

$$< u_i^2 > = \frac{1}{N_1 N_2 N_3} \cdot \sum_{n=1}^{N_1} \sum_{m=1}^{N_2} \sum_{q=1}^{N_3} \left[(\bar{u}_i)^2 + (u_i')^2 \right], \tag{2}$$

where $(\bar{u}_i)^2 = \bar{u}_i \bar{u}_i$ and $(u_i')^2 = \overline{u_i' \cdot u_i'}$.

The subgrid-scale tensor for smaller cells is

$$\tau_{ij}^u = \overline{u_i' u_j'} = -2 \cdot C_S \cdot \Delta_s^2 \cdot (2 \cdot \overline{S}_{ij}^s \cdot \overline{S}_{ij}^s)^{\frac{1}{2}} \cdot \overline{S}_{ij}^s, \tag{3}$$

where $\Delta_s = (\Delta_i \Delta_j \Delta_k)^{\frac{1}{3}}$ - is the width grid filter of the small cell.

The deformation velocity calculated in smaller cells is

$$\overline{S}_{ij}^s = \frac{1}{2} \left(\frac{\partial \bar{u}_i^s}{\partial x_j} + \frac{\partial \bar{u}_j^s}{\partial x_i} \right),$$

where $n = \overline{1, N_1}$, $m = \overline{1, N_2}$, $q = \overline{1, N_3}$

By placing expression (3) into equation (2), we can obtain the average velocity value calculated in smaller cells:

$$< u_i^2 >^s = \frac{1}{N_1 N_2 N_3} \cdot \sum_{n=1}^{N_1} \sum_{m=1}^{N_2} \sum_{q=1}^{N_3} \left[(\bar{u}_i^s)^2 - 2 \cdot C_S \cdot \Delta_s^2 \cdot (2 \cdot S_{ij}^s \cdot S_{ij}^s)^{\frac{1}{2}} S_{ij}^s \right]. \tag{4}$$

The average velocity calculated in larger cells is

$$< u_i^2 >^l = \frac{8}{N_1 N_2 N_3} \cdot \sum_{p=1}^{N_1/2} \sum_{g=1}^{N_2/2} \sum_{r=1}^{N_3/2} \left[(\bar{u}_i^l)^2 - 2 \cdot C_S \cdot \Delta_l^2 \cdot (2 \cdot S_{ij}^l \cdot S_{ij}^l)^{\frac{1}{2}} S_{ij}^l \right].$$
$$\tag{5}$$

where $\Delta_l = (\Delta_i \Delta_j \Delta_k)^{\frac{1}{3}}$ - is the width grid filter of the large cell, $\Delta_l = 2 \cdot \Delta_s$.

The deformation velocity calculated in larger cells is

$$\overline{S}_{ij}^l = \frac{1}{2} \left(\frac{\partial \bar{u}_i^l}{\partial x_j} + \frac{\partial \bar{u}_j^l}{\partial x_i} \right),$$

where $p = 1, 2, 3, ..., \frac{N_1}{2}$; $g = 1, 2, 3, ..., \frac{N_2}{2}$; $r = 1, 2, 3, ..., \frac{N_3}{2}$.

$$\bar{u}_i^l(p, g, r) = \frac{1}{8} \begin{bmatrix} \bar{u}_i^S(2p-1, 2g-1, 2r-1) + \bar{u}_i^S(2p-1, 2g, 2r-1) + \\ +\bar{u}_i^S(2p-1, 2g, 2r) + \bar{u}_i^S(2p-1, 2g-1, 2r) + \\ +\bar{u}_i^S(2p, 2g-1, 2r-1) + \bar{u}_i^S(2p, 2g, 2r-1) + \\ +\bar{u}_i^S(2p, 2g, 2r) + \bar{u}_i^S(2p, 2g-1, 2r) \end{bmatrix}.$$

We introduce the following notation:

$$F^u = \left(<\bar{u}_1^2>^s + <\bar{u}_2^2>^s + <\bar{u}_3^2>^s - <\bar{u}_1^2>^l - <\bar{u}_2^2>^l - <\bar{u}_3^2>^l \right)^2.$$

From equations (4) and (5), we can conclude

$$F^u = (Z^u - Y^u \cdot C_S)^2,$$

where

$$Z^u = \frac{1}{N_1 N_2 N_3} \cdot \sum_{n=1}^{N_1} \sum_{m=1}^{N_2} \sum_{q=1}^{N_3} (\bar{u}_i^2)^s - \frac{8}{N_1 N_2 N_3} \cdot \sum_{p=1}^{N_1/2} \sum_{g=1}^{N_2/2} \sum_{r=1}^{N_3/2} (\bar{u}_i^2)^l,$$

$$Y^u = \frac{1}{N_1 N_2 N_3} \cdot \sum_{n=1}^{N_1} \sum_{m=1}^{N_2} \sum_{q=1}^{N_3} (-2 (\Delta_s)^2 (2 S_{ij}^s S_{ij}^s)^{\frac{1}{2}} S_{ij}^s) -$$
$$- \frac{8}{N_1 N_2 N_3} \cdot \sum_{p=1}^{N_1/2} \sum_{g=1}^{N_2/2} \sum_{r=1}^{N_3/2} (-2 (\Delta_l)^2 (2 S_{ij}^l S_{ij}^l)^{\frac{1}{2}} S_{ij}^l).$$

The condition for achieving the minimum is

$$\frac{\partial F^u}{\partial C_S} = -2 (Z^u - Y^u \cdot C_S) \cdot Y^u = 0.$$

Thus, $Z^u - Y^u \cdot C_S = 0$.

At a certain time layer T_{step} the empirical coefficient of viscosity model is calculated by the following formula: $C_S = Z^u / Y^u$, where $T_{step} = 10 \cdot \tau$, τ-time step.

4 Method for Calculating the Small-Scale Magnetic Field

Here is used the same grid, which was used to calculate the small-scale turbulence coefficient, which deals with the grid twice the size of cells along each axis.

The average values of magnetic field strength H_1^2, H_2^2, H_3^2 for the total volume of the calculated area of the liquid flow are marked $\langle H_1 \rangle^2$, $\langle H_2 \rangle^2$, $\langle H_3 \rangle^2$. These values can be calculated using smaller and larger calculation grids:

$$< H_i^2 > = \frac{1}{N_1 N_2 N_3} \cdot \sum_{n=1}^{N_1} \sum_{m=1}^{N_2} \sum_{q=1}^{N_3} [(\bar{H}_i)^2 + (H_i')^2], \tag{6}$$

where $(\bar{H}_i)^2 = \bar{H}_i \bar{H}_i$ and $(H_i')^2 = \overline{H_i' \cdot H_i'}$.

The magnetic subgrid-scale tensor for smaller cells is

$$\tau_{ij}^H = \overline{H_i' H_j'} = -2 \cdot D_S \cdot \Delta_s^2 \cdot (2 \cdot \bar{J}_{ij}^s \cdot \bar{J}_{ij}^s)^{\frac{1}{2}} \cdot \bar{J}_{ij}^s. \tag{7}$$

The magnetic rotation tensor calculated in smaller cells is

$$\bar{J}_{ij}^s = \frac{1}{2} \left(\frac{\partial \bar{H}_i^s}{\partial x_j} - \frac{\partial \bar{H}_j^s}{\partial x_i} \right),$$

where $n = \overline{1, N_1}, \ \ m = \overline{1, N_2}, \ \ q = \overline{1, N_3}$

By placing expression (7) into equation (6), we can obtain the average velocity value calculated in smaller cells:

$$< H_i^2 >^s = \frac{1}{N_1 N_2 N_3} \cdot \sum_{n=1}^{N_1} \sum_{m=1}^{N_2} \sum_{q=1}^{N_3} \left[(\overline{H}_i^s)^2 - 2 \cdot D_S \cdot \Delta_s^2 \cdot (2 \cdot J_{ij}^s \cdot J_{ij}^s)^{\frac{1}{2}} J_{ij}^s \right]. \quad (8)$$

The average value of magnetic field strength calculated in larger cells is

$$< H_i^2 >^l = \frac{8}{N_1 N_2 N_3} \cdot \sum_{p=1}^{N_1/2} \sum_{g=1}^{N_2/2} \sum_{r=1}^{N_3/2} \left[(\overline{H}_i^l)^2 - 2 \cdot D_S \cdot \Delta_l^2 \cdot (2 \cdot J_{ij}^l \cdot J_{ij}^l)^{\frac{1}{2}} J_{ij}^l \right].$$
$$(9)$$

The magnetic rotation tensor calculated in larger cells is

$$\overline{J}_{ij}^l = \frac{1}{2} \left(\frac{\partial \bar{H}_i^l}{\partial x_j} - \frac{\partial \bar{H}_j^l}{\partial x_i} \right),$$

where $p = 1, 2, 3, ..., N_1/2; \ g = 1, 2, 3, ..., N_2/2; \ r = 1, 2, 3, ..., N_3/2.$

$$\bar{H}_i^l(p, g, r) = \frac{1}{8} \left[\begin{array}{l} \bar{H}_i^S(2p - 1, \ 2g - 1, \ 2r - 1) + \bar{H}_i^S(2p - 1, \ 2g, \ 2r - 1) + \\ + \bar{H}_i^S(2p - 1, \ 2g, \ 2r) + \bar{H}_i^S(2p - 1, \ 2g - 1, \ 2r) + \\ + \bar{H}_i^S(2p, \ 2g - 1, \ 2r - 1) + \bar{H}_i^S(2p, \ 2g, \ 2r - 1) + \\ + \bar{H}_i^S(2p, \ 2g, \ 2r) + \bar{H}_i^S(2p, \ 2g - 1, \ 2r) \end{array} \right].$$

We introduce the following notation:

$$F^H = \left(< \overline{H}_1^2 >^s + < \overline{H}_2^2 >^s + < \overline{H}_3^2 >^s - < \overline{H}_1^2 >^l - < \overline{H}_2^2 >^l - < \overline{H}_3^2 >^l \right)^2.$$

From equations (8) and (9), we can conclude

$$F^H = (Z^H - Y^H \cdot D_S)^2,$$

where

$$Z^H = \frac{1}{N_1 N_2 N_3} \cdot \sum_{n=1}^{N_1} \sum_{m=1}^{N_2} \sum_{q=1}^{N_3} (\overline{H}_i^2)^s - \frac{8}{N_1 N_2 N_3} \cdot \sum_{p=1}^{N_1/2} \sum_{g=1}^{N_2/2} \sum_{r=1}^{N_3/2} (\overline{H}_i^2)^l,$$

$$Y^H = \frac{1}{N_1 N_2 N_3} \cdot \sum_{n=1}^{N_1} \sum_{m=1}^{N_2} \sum_{q=1}^{N_3} \left(-2 \left(\Delta_s \right)^2 (2 J_{ij}^s J_{ij}^s)^{\frac{1}{2}} J_{ij}^s \right)$$
$$- \frac{8}{N_1 N_2 N_3} \cdot \sum_{p=1}^{N_1/2} \sum_{g=1}^{N_2/2} \sum_{r=1}^{N_3/2} \left(-2 \left(\Delta_l \right)^2 (2 J_{ij}^l J_{ij}^l)^{\frac{1}{2}} J_{ij}^l \right).$$

The condition for achieving the minimum is

$$\frac{\partial F^H}{\partial D_S} = -2\left(Z^H - Y^H \cdot D_S\right) \cdot Y^H = 0.$$

Hence, $Z^H - Y^H \cdot D_S = 0$.

Thus, the empirical coefficient of viscosity model for magnetic field at a certain time step T_{step} assumes the following form: $D_S = Z^H/Y^H$.

5   Numerical Method

To solve the problem of homogeneous incompressible MHD turbulence, a scheme of splitting by physical parameters is used:

I. $(\boldsymbol{u}^* - \boldsymbol{u}^n)/\tau = -\left(\boldsymbol{u}^n\nabla\right)\boldsymbol{u}^* + A\left(\boldsymbol{H}^n\nabla\right)\boldsymbol{H}^n + (1/Re)\left(\Delta\boldsymbol{u}^*\right) - \nabla\tau^u,$

II. $\Delta p = \nabla\boldsymbol{u}^*/\tau,$

III. $\left(\boldsymbol{u}^{n+1} - \boldsymbol{u}^*\right)/\tau = -\nabla p.$

IV. $\left(\boldsymbol{H}^{n+1} - \boldsymbol{H}^n\right)/\tau = -rot(\boldsymbol{u}^{n+1} \times \boldsymbol{H}^{n+1}) + \nu_m\Delta\boldsymbol{H}^{n+1} - \nabla\tau^H.$

The following physical interpretation of the splitting diagram is suggested. During the first stage, the Navier-Stokes equation is solved without the pressure consideration. For the approximation of convective and diffusion equation members, a compact scheme of an increased order of accuracy is used [13]. During the second stage, the Poisson equation is solved, which is obtained from the continuity equation by considering the velocity fields of the first stage. For the three-dimensional Poisson equation, an original solution algorithm was developed – a spectral transform in combination with the matrix run. During the third stage, the obtained pressure field is used to recalculate the final velocity field. During the fourth stage, the obtained velocity field is used to solve the equation to obtain the components of the magnetic field strength, which are included in the initial equation.

6   Numerical Modeling Results

Numerical model allows to describe the homogeneous magnetohydrodynamic turbulence decay based on large eddy simulation. For this task, the kinematic viscosity $\nu = 10^{-4}$ was taken constant and the magnetic viscosity were set in the range of $\nu_m = 10^{-3} \div 10^{-4}$. The characteristic values of the velocity, length, magnetic field strength were taken equal to: $U_{CH} = 1$, $L_{CH} = 1$, $H_{CH} = 1$ respectively. Reynolds number is $Re = 10^4$, the magnetic Reynolds number varied depending on the magnetic viscosity coefficient. The Alfven number characterizing the motion of conductive fluid for various numbers of magnetic Reynolds:

$A = Ha^2/Re_m$, where Hartmann number is $Ha = 1$. For the calculations used grid size 128x128x128. The time step was taken equal $\Delta\tau = 0.001$.

As result of simulation at different magnetic Reynolds numbers were obtained the following turbulence characteristics: kinetic energy, magnetic energy, integral scale longitudinal correlation functions.

Figure 1 shows the evolution of the kinetic and magnetic energy changes depending on Re_m number at different points in time. Re_m was selected in range $10^3 \div 10^4$. For the first time, the result is obtained for the turbulence decay modeling under the impact of a magnetic field caused by the change in Re_m number on the kinetic energy of the turbulent flow of a fluid with various conductive properties. It is easily seen the kinetic energy in case of a high environment conductivity, when $Re_m = 10^3$, the friction force increases and the flow velocity is reduced more quickly in case of a high environment conductivity than when $Re_m = 10^4$, which is consistent with a low conductivity environment, in this option, the friction force has less impact on the flow rate. Thus, 1 illustrates the dynamics of the mutual influence of magnetic and kinetic energies at different points in time: at the initial point in time, the kinetic and magnetic energies are defined identically; at the next point when the fluid with a higher conductivity is studied, the turbulence decay occurs faster than in case where Re_m starts to increase, which determines the fluid with a lesser conductivity. When value $Re_m = 10000$, the turbulence decay virtually corresponds to the case of an isotropic turbulence decay, as per Abdibekov and Zhakebayev [14].

According to semi-empirical theory of turbulence integral scale should grow with time. The results presented in Figure 2 illustrates the effect of magnetic viscosity on the internal structure of the MHD turbulence. Variation of the coefficient of magnetic viscosity leads to a proportional change in the integral scale. Figure 2 shows that the size of large eddies rapidly increases at small number of magnetic Reynolds $Re_m = 10^3$, than in the case, when $Re_m = 10^4$ which leads to fast energy dissipation.

Figure 3 shows the change in the micro scale - calculated at different numbers of magnetic Reynolds 1) $Re_m = 10^3$; 2) $Re_m = 2 \cdot 10^3$; 3) $Re_m = 5 \cdot 10^3$; 4) $Re_m = 10^4$. Figure 3 shows the change of the Taylor microscale at different magnetic Reynolds numbers. It can be seen that in the case $Re_m = 10^3$ when the magnetic viscosity coefficient is large then the dissipation rate increases. In the case when the magnetic viscosity coefficient is smaller then the scale gradually increases, and the small scale structure of the turbulence tends to slowly isotropy. This also indicates that with small numbers Re_m the decay of isotropic turbulence occurs faster than in the case when Re_m is high.

Figure 4 shows the changes of the longitudinal correlation function calculated at $Re_m = 10^3$ and $Re_m = 10^4$. These illustrations also show that there are an influence of the magnetic field on the isotropic turbulence decay, as these figures are fixed the result of changes in the correlation functions at different Re_m.

The correlation function is expressed the average by volume the correlation ratio between the components of the velocity at various points, the farther points are located between the various components of the velocity, the smaller should

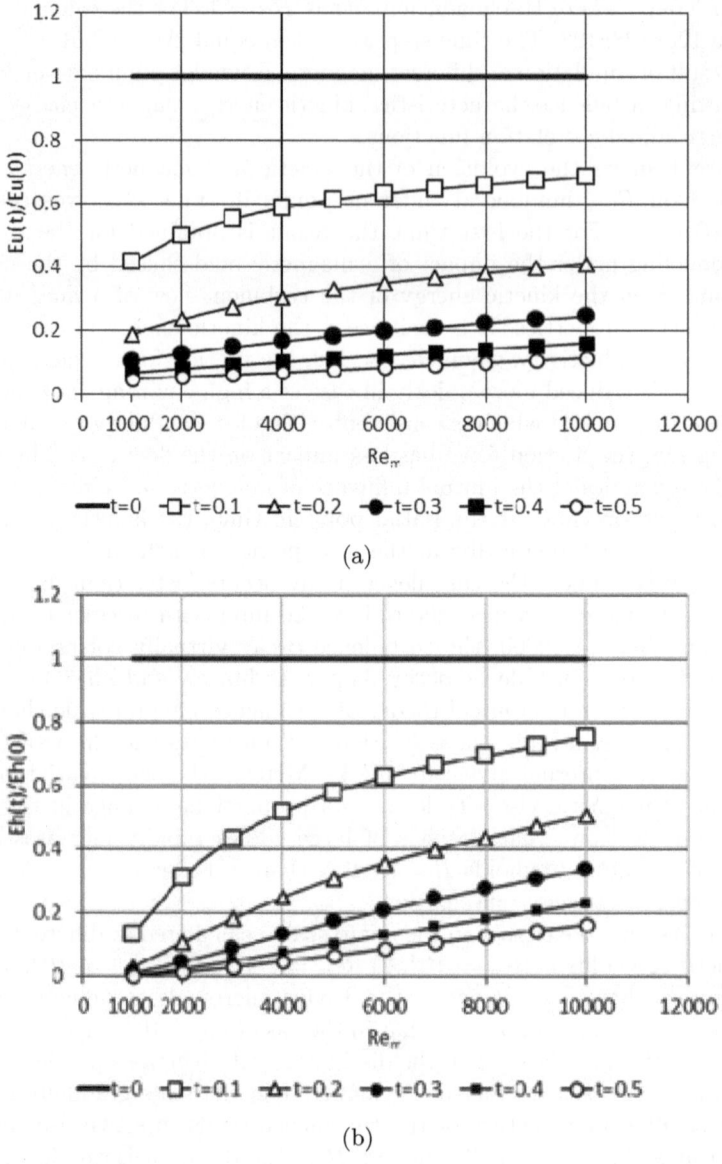

Fig. 1. Change of the kinetic (a) and magnetic (b) energies depending on the Re_m number at different points in time

be the correlation coefficients, i.e. they should be close to zero. Figure 4a shows the change in the longitudinal correlation function $f(r)$ in time and calculated at $Re = 10^4$, $Re_m = 10^3$. It is seen that with increasing value r of the function

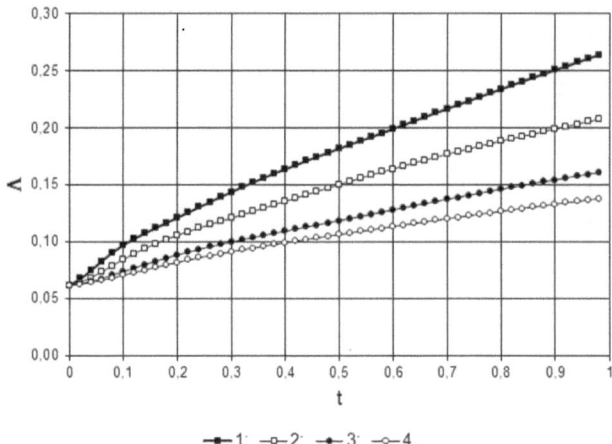

Fig. 2. Change of the integral turbulence scale calculated at different magnetic Reynolds numbers: 1) $Re_m = 10^3$; 2) $Re_m = 2 \cdot 10^3$; 3) $Re_m = 5 \cdot 10^3$; 4) $Re_m = 10^4$

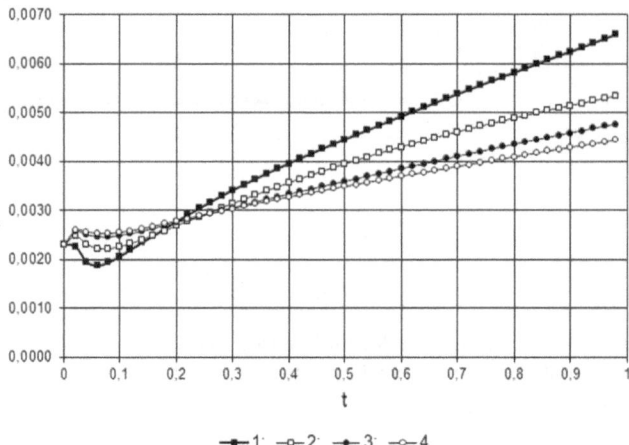

Fig. 3. Change of Taylor-scale calculated at different magnetic Reynolds numbers: 1) $Re_m = 10^3$; 2) $Re_m = 2 \cdot 10^3$; 3) $Re_m = 5 \cdot 10^3$; 4) $Re_m = 10^4$

tends to zero. Character of the correlations change corresponds to the change of the correlation functions given in [14].

From the figures it is seen that in the case of high medium conductivity at $Re_m = 10^3$ the frictional force increases and the flow rate is reduced faster than, at $Re_m = 10^4$, that corresponds to the low conductivity of the medium, in this version, the frictional force have minimal impact on the flow velocity. Based on the study of the results determined that the first part of the turbulent kinetic energy is used for turbulent mixing, the second part - at creating magnetic field and the third part - on the forces of resistance between the components of the velocity and magnetic tension.

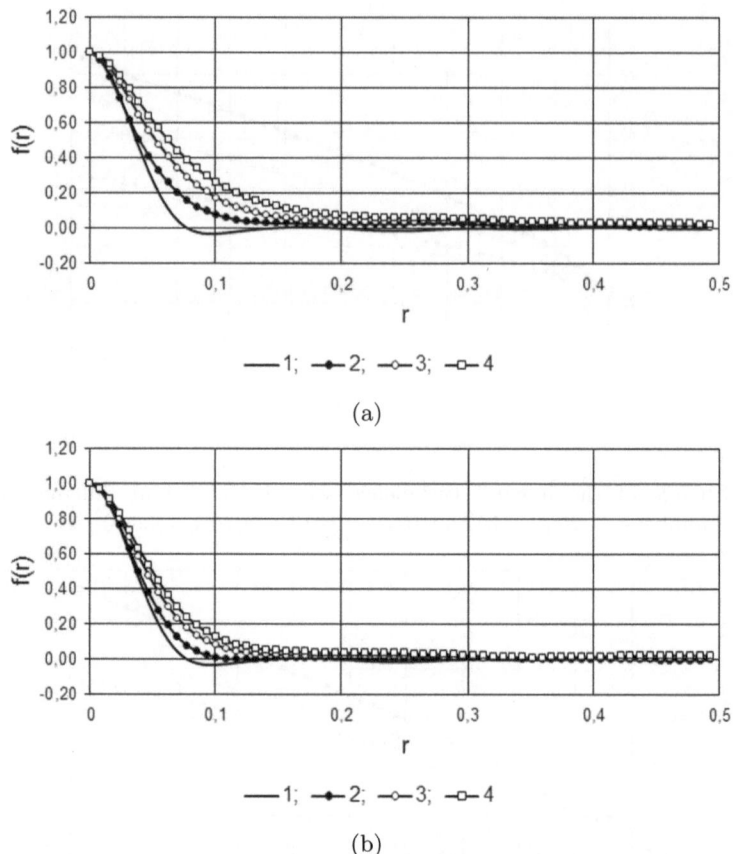

Fig. 4. Change the longitudinal correlation function $f(r)$ when (a) $Re_m = 10^3$ and (b) $Re_m = 10^4$ at different points in time: 1) $t = 0$; 2) $t = 0.2$; 3) $t = 0.3$; 4) $t = 0.5$

7   Conclusions

Based on the method large-eddy simulation was produced the numerical modelling of influence magnetic viscosity to decay of homogeneous magnetohydrodynamic turbulence, analyzing simulation results it is possible to make the following conclusion: the magnetic viscosity of the flow has a significant influence on the MHD turbulence, and therefore can be used for process control in the preparation semiconductor structures of single crystals. Obtained results allow sufficiently accurately calculate the change characteristics of homogeneous magnetohydrodynamic turbulence over time at large magnetic Reynolds numbers. Thus, the numerical algorithm was developed for solving unsteady three-dimensional magnetohydrodynamic equations, for modeling MHD turbulence decay at different magnetic Reynolds numbers. Physical processes and phenomena of homogeneous magnetohydrodynamic turbulence identified in the numerical simulation. The proposed method can be used to solve the MHD turbulence without significant changes.

References

1. Batchelor, G.K.: On the spontaneous magnetic field in a conducting liquid in turbulent motion. Proc. Roy. Soc. A **201**(16), 405–416 (1950)
2. Schumann, U.: Numerical simulation of the transition from three- to two-dimensional turbulence under a uniform magnetic field. J. Fluid Mech. **74**, 31–58 (1976)
3. Moffatt, H.K.: On the suppression of turbulence by a uniform magnetic field. J. Fluid Mech. **28**, 571–592 (1967)
4. Hossain, M.: Inverse energy cascades in three dimensional turbulence. Phys. Fluids B. **3**(2), 511–514 (1991)
5. Zikanov, O., Thess, A.: Direct numerical simulation of forced MHD turbulence at low magnetic Reynolds number. J. Fluid Mech. **358**(1), 299–333 (1998)
6. Vorobev, A., Zikanov, O., Davidson, P.A., Knaepen, B.: Anisotropy of magnetohydrodynamic turbulence at low magnetic Reynolds number. Phys. Fluid 17 (2005)
7. Burattini, P., Zikanov, O., Knaepen, B.: Decay of magnetohydrodynamic turbulence at low magnetic Reynolds number. J. Fluid Mech. **657**, 502–538 (2010)
8. Knaepen, B., Kassinos, S., Carati, D.: Magnetohydrodynamic turbulence at moderate magnetic Reynolds number. J. Fluid Mech. **513**(3), 199–220 (2004)
9. Knaepen, B., Moin, P.: Large-eddy simulation of conductive flows at low magnetic Reynolds number. Physics of Fluids **16**, 1255–1261 (2004)
10. Sahoo, G., Perlekar, P., Panditn, R.: Systematics of the magnetic-Prandtl-number dependence of homogeneous, isotropic magnetohydrodynamic turbulence. New J. Phys. **13**, 1367–2630 (2011)
11. Ievlev, V.M.: The method of fractional steps for solution of problems of mathematical physics. Science Nauka, Moscow (1975)
12. Sirovich, L., Smith, L., Yakhot, V.: Energy spectrum of homogeneous and isotropic turbulence in far dissipation range. Physical Review Letters **72**(3), 344–347 (1994)
13. Zhumagulov, B., Abdibekov, U., Zhakebaev, D., Zhubat, K.: Modelling isotropic turbulence decay based on the LES. Mathematical modelling **25**(1), 18–32 (2013)
14. Abdibekov, U.S., Zhakebaev, D.B.: Modelling of the decay of isotropic turbulence by the LES. J. Phys.: Conf. Ser. 318 (2011)

Enhancement of the In-Situ Leach Mineral Mining Process by the Hydrodynamic Method

Karlygash Alibayeva$^{(\boxtimes)}$ and Aidarkhan Kaltayev

Al-Farabi Kazakh National University, Al-Farabi ave. 71, 050040 Almaty, Kazakhstan
{Karlygash.Alibaeva,Aidarkhan.Kaltayev}@kaznu.kz

Abstract. In this paper, a hydrodynamic method of enhancement of mineral extraction is numerically studied. Results of the preliminary performed computing and experimental data show that during the mineral extraction process a stagnation zone is formed in layer. Formation of the stagnation zone is caused by the absence of reagent flow in it. Such zones result in reduction of the degree of the deposit development. In this connection, there is a need to conduct research on improving the mineral extraction degree by controlling a seepage in a layer. Therefore, the hydrodynamic method is used to engage stagnation zones into the leaching process. The hydrodynamic method of enhancement based on changing reagent flow direction during the in-situ leach process by reversing the wells.

Keywords: In-situ leaching (ISL) · Mineral extraction degree · Stagnation zone · Hydrodynamic method · Reversing well

1 Introduction

In-situ leaching (ISL) is a method for development of ore deposits without lifting the ore to the surface. This method is performed by drilling of wells through the mineral ore bodies, supply of solution into mineral ore bodies, lifting of mineral bearing solutions to the surface. The ISL method is used for mining low concentrated and deep-laying mineral deposits such as uranium, copper and gold.

The results of computation show that filtration of solution almost does not exist between production wells during the ISL mineral mining. Therefore so-called stagnation zone is appearing, which leads to decrease an extraction degree of mineral. One of the solution of avoiding this problem is using wells reversely (change production well over to injection and vice-versa) in stagnation zones. This method is called as hydrodynamic method of enhancement, and filtration flow direction is changed by reversing the technological wells.

In this work as a mineral is considered uranium, and as a reagent − sulfuric acid. The design of ISL well fields varies greatly depending on the local conditions such as permeability, sand thickness, deposit type, ore grade and distribution. Therefore, numerical study is carried out for linear as well as for hexagonal schemes of wells location.

© Springer International Publishing Switzerland 2015
N. Danaev et al. (Eds.): CITech 2015, CCIS 549, pp. 26–32, 2015.
DOI: 10.1007/978-3-319-25058-8_3

2   Mathematical Model

Transfer reaction of useful element from solid phase to liquid phase is produced as

$$\nu_m M + \nu_r R = \nu_p P + \nu_w W \tag{1}$$

Where M - denotes the gram-molecule of mineral (uranium) in solid phase, R - gram-molecule of reactant (sulfuric acid), P - gram-molecule of useful element of dissolved uranium, W - gram-molecule of by-product in liquid phase (ex., water), ν - stoichiometric coefficient, where subscripts m, r, p and w denote reactant, mineral, useful element and water, respectively.

Governing equations describing the ISL process are mass conservation law, the Darcy law and conservation equation of mineral in solid phase, reagent (sulfuric acid), mineral in liquid phase:

$$div(KgradH) + \sum_{i=1}^{n} q_{si}\delta(\boldsymbol{x} - \boldsymbol{x}_i) = 0 \tag{2}$$

$$\boldsymbol{V} = -KgradH \tag{3}$$

$$\frac{\partial C_m}{\partial t} = -\beta\theta C_r C_m \tag{4}$$

$$\frac{\partial\theta C_r}{\partial t} = div(\theta DgradC_r) - \boldsymbol{V}grad(\theta C_r) - \nu_1\beta\theta C_m C_r + \sum_{i=1}^{n} q_i C_r^0\delta(\boldsymbol{x} - \boldsymbol{x}_i) \tag{5}$$

$$\frac{\partial\theta C_p}{\partial t} = div(\theta DgradC_p) - \boldsymbol{V}grad(\theta C_p) + \nu_2\beta\theta C_m C_r - \sum_{i=1}^{n} q_i C_p\delta(\boldsymbol{x} - \boldsymbol{x}_i) \tag{6}$$

Here $\nu_1 = \frac{\nu_r R}{\nu_m M}$, $\nu_2 = \frac{\nu_p P}{\nu_m M}$, K - is the permeability coefficient, H - is the head pressure, $\boldsymbol{V}$ - is the filtration rate, C_m - is the concentration of mineral in solid phase, C_m^0 - is the initial content of mineral in layer, C_r - is the concentration of sulfuric acid in solution, C_r^0 - is the concentration of reagent on injection well, C_p - is the concentration of useful element (uranium) in solution, θ - is the porosity of layer, β - is the coefficient, characterizing reaction rate, $\delta(\boldsymbol{x} - \boldsymbol{x}_i)$ - is the Dirac delta function by which location of the well is given, $\boldsymbol{x}_i$ - coordinates of wells, q - is the debit of well ($q < 0$ for production well, $q > 0$ for inject well). D - is the hydrodynamic dispersion coefficient [2].

2.1   Initial and Boundary Condition

Equations (4) - (6) are solved at initial and boundary conditions. At initial time mineral distribution C_m in layer is known, C_r concentration of reagent and dissolved useful element C_p does not exists. Therefore, the initial condition is given as

$$C_m|_{t=0} = C_m^0; \quad C_r|_{t=0} = 0; \quad C_p|_{t=0} = 0 \qquad (7)$$

Due to the symmetry of the considering area it is sufficient to simulate only the symmetric part. In that case, Neumann condition is used on the symmetric boundary

$$\frac{\partial C_m}{\partial x}\Big|_G = 0; \quad \frac{\partial C_m}{\partial y}\Big|_G = 0; \quad \frac{\partial C_m}{\partial z}\Big|_G = 0;$$
$$\frac{\partial C_r}{\partial x}\Big|_G = 0; \quad \frac{\partial C_r}{\partial y}\Big|_G = 0; \quad \frac{\partial C_r}{\partial z}\Big|_G = 0; \qquad (8)$$
$$\frac{\partial C_p}{\partial x}\Big|_G = 0; \quad \frac{\partial C_p}{\partial y}\Big|_G = 0; \quad \frac{\partial C_p}{\partial z}\Big|_G = 0;$$

On the border of the deposit it is possible to give value of concentrations, and in that case the Dirichlet boundary conditions are used

$$C_m|_G = C_M; \quad C_r|_G = C_R; \quad C_p|_G = C_P \qquad (9)$$

3   Numerical Model

The differential equation for head pressure (2) is solved by over-relaxation iterative method [4]; filtration rate is calculated from Darcy law by using defined values of hydraulic head. Transport equations of reagent concentration in liquid phase (5), useful element concentration in solid phase (4), and its transition to liquid phase (6) are solved together by the implicit Crank-Nicolson scheme. Crank-Nicolson scheme is implemented in three stages in case of 3D problem by using splitting technique of the alternating direction implicit (ADI) method [4].

4   Results of Calculation

4.1   Results of Calculation for Linear Well Location

Technology of reversing wells which allows to change the direction of streamlines in the formation is applied to extract the mineral from stagnation (dead) zones.

Two options of reversing wells are considered: when the main quantity of mineral is extracted I) the action of all injection wells is stopped and some of the production wells is used as an injection to direct reagent flow towards the stagnation zone; II) all production wells changed over to injection and vice-versa.

Calculation showed that capture area of stagnation zone differs at various time. So the reversing a well at three different time are considered to get an optimal value of extraction degree: a) $T = 100$ days, b) $T = 200$ days, c) $T = 300$ days.

Dependence of mineral extraction degree and mineral concentration on production well on time are represented in the Fig. 1 at the reversing wells according to the option I and in the Fig. 2 at the reversing wells according to the option II.

Comparative results of the reversing the wells according to the option I and option II are shown in the Fig.3. The curve with squares refers to the value of extraction degree without reversing the wells; the curve with triangles refers to

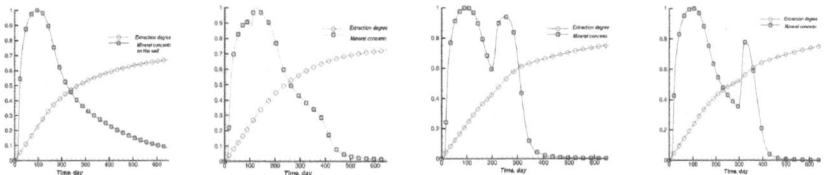

Fig. 1. Dependence of extraction degree of layer and mineral concentration on production well on time: a) without reversing the well; b) reversing the well at $T = 100$ days; c) reversing the well at $T = 200$ days; d) reversing the well at $T = 300$ days. Curve with circles - extraction degree of layer; curve with squares - mineral concentration on production well.

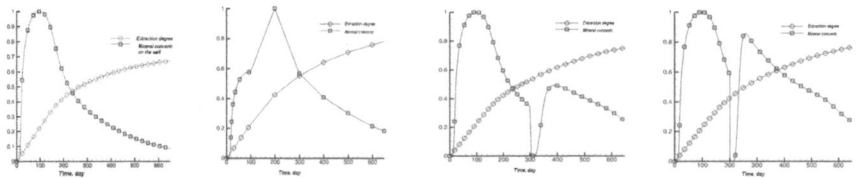

Fig. 2. Dependence of extraction degree of layer and mineral concentration on production well on time: a) without reversing the well; b) reversing the well at T=100 days; c) reversing the well at T=200 days; d) reversing the well at T=300 days. Curve with circles - extraction degree of layer; curve with squares - mineral concentration on production well.

the value of extraction degree in which reversing the well is used at T=100 days; curve with asterisks - at T=200 days; curve with circles - at T=300 days. The obtained curves show that the using wells reversely in stagnation zone leads to increase the extraction degree. In case of option I extraction degree increases for 8% and in case of option II it increases for 11% (Table 1).

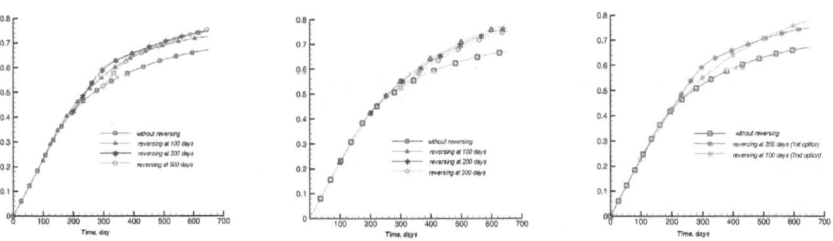

Fig. 3. Comparison of extraction degrees a)option I, b) option II, c) maximal values of the option I and option II.

Table 1. Comparative analysis of the calculation results for the options I and II (Linear well)

	Reversing time (day)	Extraction degree (%)	The mineral content in productive solutions at the production well (gr/l)	Deposit development time (day)
	Without reversing	67	0,084	650
Option I	at =100	72	0,012	650
	at =200	75	0,002	650
	at =300	75	0,001	650
Option II	at =100	78	0,3	650
	at =200	76	0,28	650
	at =300	75	0,25	650

4.2   Results of Calculation for Hexagonal Well Pattern

During the in-situ mineral leaching in the case of a hexagonal well location the results of calculations and experimental data show the presence of stagnation (dead) zones where there is no filtering solution (between the petals of the hexagon). Two kind of schemes for reversing the hexagonal well locations are represented in Figure 4. The case without reversing wells and two options of reversing flow directions are considered. In option I all injection wells' action is stopped and half of the remained production wells are reversed to the injection wells forming a linear wells pattern (Figure 4, b), and in option II the flow direction of all wells are changed by reversing all the wells (Figure 4, c).

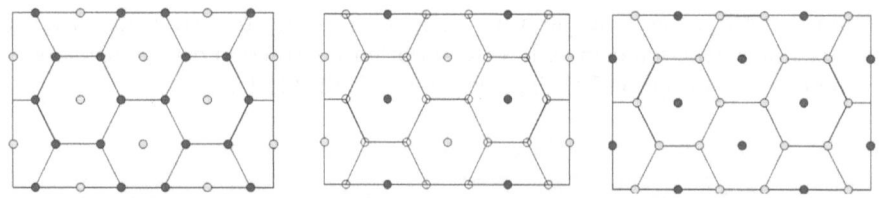

Fig. 4. Various scheme of reversing wells for the hexagonal well location.

In Figure 5 the distribution of minerals in the solid phase is represented at various time. Similarly to the linear well pattern the study of reversing wells for the hexagonal case is conducted at three various time: (a) T = 100 days, (b) T = 150 days, (c) T = 200 days. The dependence of mineral extraction degree on time is represented in Figure 6 for the option I (Figure 6, a) and II (Figure 6, b). Where the curve with circles corresponds to the case without reversing the flow direction in the well, curve with rhombus corresponds to reversing wells at

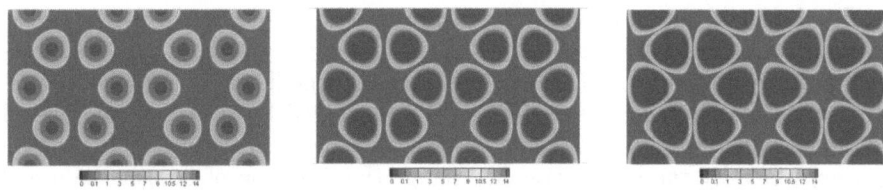

Fig. 5. Solid mineral distribution for the hexagonal well location.

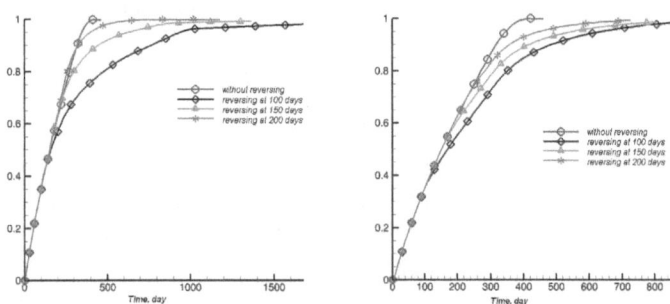

Fig. 6. Comparison of mineral extraction degree: a) option I, b) option II.

Table 2. Comparative analysis of the numerical results for the options I and II (Hexagonal well)

	Reversing time (day)	Extraction degree (%)	The mineral content in productive solutions at the production well (gr/l)	Deposit development time (day)
	Without reversing	99	0,045	462
Option I	at =100	98	0,043	1780
	at =150	99	0,043	1600
	at =200	99	0,044	1400
Option II	at =100	99	0,044	972
	at =150	99	0,045	903
	at =200	99	0,043	815

T = 100 days, curve with the triangles - at T = 150 days, curve with asterisks - at T = 200 days.

  Calculation results show that the applying reversing technology for the hexagonal wells pattern is unreasonable. As far as to extract a maximal mineral in case of reversing the wells it needs more time than the case of without reversing. Comparative analysis is represented in Table 2.

The large distance between the production and injection wells after the reversing wells leads to a reduction of the pressure gradient, and this negatively affects the filtration rate of the solution. Inexpediency of applying reversing technology according to the option I is explained by the above-mentioned factors. Inexpediency of applying reversing technology according to the option II is explained by the loss of time to change the direction of flow of the useful component accumulated away from the new production wells. This implies, an application the technology of reversing in the internal zones of hexagonal scheme where stagnation zone is formed between the production wells is inefficient. However, this technology can be applied near the boundary areas of deposits, which have complex geometric shape.

5   Conclusion

Considered problem is an actual from the point of view of mineral mining technology. During the mineral extraction process, a stagnation zone is formed in layer and an extraction degree of mineral cannot reach to the expecting value. To solve this problem it is suggested to use wells reversely (hydrodynamic method) in stagnation zones. However, it needs to know at which time wells should be use reversely in order to optimize the process. Thereby several options are considered in this work. Obtained numerical results show that using hydrodynamic enhancement method in stagnation zone enables to increase the extraction degree. Depending on the considering options of well reversing, the extraction degree is increased from 8% to 11% in case of linear well location. In addition, results of calculation showed that using reversing wells in the inner zones of hexagon is inefficient. This technology of enhancement can also be applied in boundary areas of deposits, which have complex geometric shape. Developed 3D numerical model is used for investigation uranium extraction process by the in-situ leaching (ISL) method, however, it can be easily applied for study other minerals extracting by ISL process.

References

1. Manual of acid in situ leach Uranium Mining Technology. IAEA, Vienna (2001). IAEA-TECDOC-1239. ISSN 1011–4289
2. Zheng, C., Wang, P.P.: MT3DMS: A Modular Three-Dimensional Multispecies Transport Model for Simulation of Advection, Dispersion, and Chemical Reaction of Contaminants in Groundwater Systems. Documentation and User's Guide. Contract Report SERDP-99-1, U.S. Army Engineer Research and Development Center. Vicksburg (1999)
3. Bommer, P.M., Schechter, R.S.: Mathematical Modelling of in-situ uranium leaching. J. Soc. Petrol Eng. **19**(7), 393–400 (1979)
4. Flecher, K.: Numerical methods in fluid dynamics, Moscow (1991)
5. Boitsov, V.E., Vercheba, A.A.: Geologo-promyshlennyie tipy mestorozhdenii urana, Moscow (2008). (in Russian)
6. Turayev, N. S., Zherin, Y.Y.: Khimiya i tekhnologiya urana, Moscow (2005). (in Russian)

Numerical Modeling of Artificial Heart Valve

Dmitriy Dolgov[✉] and Yury Zakharov

Kemerovo State University, Kemerovo, Russia
9erthalion6@gmail.com, zaxarovyn@rambler.ru

Abstract. The paper is dedicated to the mathematical model describing dynamics of an artificial heart valve being moved by inhomogeneous incompressible fluid flow with variable viscosity, and its computational method. The modeling results of tricuspid valve performance are presented.

Keywords: Viscous inhomogeneous fluid · Artificial heart valve · Immersed boundary method

1 Introduction

The importance of medical researches of human blood circulatory system can hardly be overestimated, because this kind of knowledge is extremely practical and significant. Annually approximately 250 000 surgeries are performed in the world to restore or replace damaged heart valves [1], and the quantity is expected to increase [2]. The solution of scientific and technical problems of the artificial valves creation depends on correct understanding of the interaction between blood flow and valve leaflets. Mathematical modeling of artificial heart valves performance enables to get thorough understanding of its internal processes in order to improve its design. There are many researches devoted to the mathematical and numerical modeling of heart valve performance, based on which two main problem solution approaches were defined.

First approach is related to the finite element methods ([3], [4], [5]). They enable to take into consideration the complex geometry of heart, bu the necessity to take into account the interaction between fluid and flexible walls requires constant rebuilding of the computational grid to meet the changing geometry of the object of research. It appears to be time and computational resources consuming.

The second approach, which is related to the immersed boundary method, is under discussion in this paper ([6], [7], [8], [9]). It can be used for the problems with complex geometry, and it doesn't require grid modification.

There are various improvements of this method, in order to model more and more complex problems. In the research [10] a formulation of this method was proposed for the three dimensional flow problem of two non mixed (separated by flexible barriers) fluids of different viscosity and density. In the papers [11], [12] this method application in case of the two dimensional problem of two component fluid flow is presented.

We propose to describe the blood flow in the flexible large blood vessels and the artificial heart valve as a three dimensional nonstationary flow of viscous incompressible fluid with variable viscosity and density (see [13], [14], [15], [16]). Thus, the

© Springer International Publishing Switzerland 2015
N. Danaev et al. (Eds.): CITech 2015, CCIS 549, pp. 33–43, 2015.
DOI: 10.1007/978-3-319-25058-8_4

goal of this work is to develop a mathematical model and a solution method of the problem of artificial heart leaflet dynamics inside a blood vessel taking into account the inhomogeneous structure of the blood, and admixture (formed elements) circulation inside a blood vessel.

2   Formulation of the Problem

We consider a nonstationary problem of blood flow inside a vessel with a valve. Blood consists of plasma and formed elements, which are approximately 45 % of the entire volume [17]. Vessel walls and valve leaflets consist of a large number of thin collagen fibers, they are flexible and can change their form depending on the fluid flow. For example, the tricuspid aortic valve, lies between the left ventricle and the aorta, and prevents blood backflow (see Fig. 1):

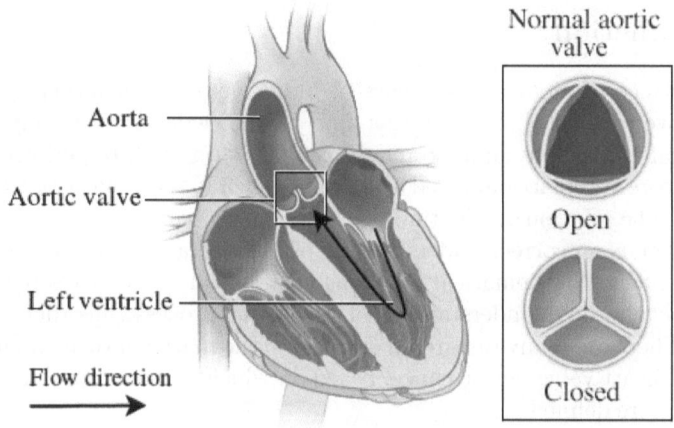

Fig. 1. Aortic valve and its location inside heart

We model the blood as a viscous incompressible inhomogeneous two component fluid with variable viscosity, and vessel wall and valve leaflets as a fluid impermeable surface with specified stiffness. Vessel and valve leaflets are deformed under the fluid pressure.

Since blood circulates through vessels under the pressure created by cardiac beats then the problem of the blood flow can be described by Navier-Stokes non-stationary system of differential equations [13]:

$$\frac{\partial \boldsymbol{u}}{\partial t} + (\boldsymbol{u} \cdot \nabla)\boldsymbol{u} = -\frac{1}{\rho}\nabla p + \nabla \sigma + \boldsymbol{f} \tag{1}$$

$$\frac{\partial \rho}{\partial t} + \nabla \cdot (\rho \boldsymbol{u}) = 0 \tag{2}$$

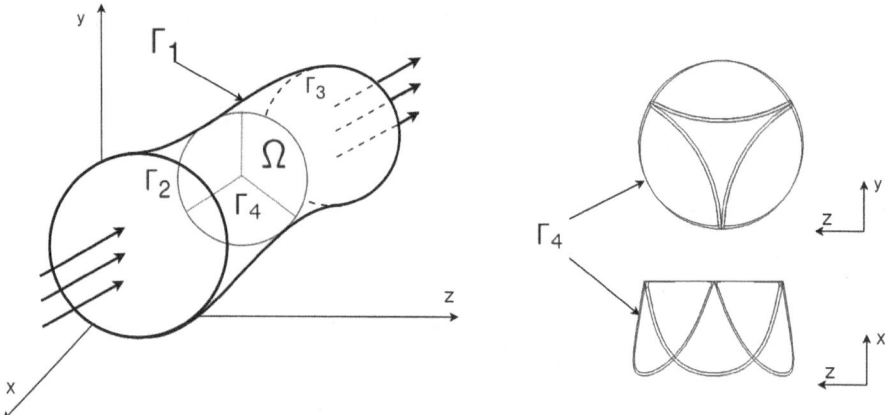

Fig. 2. Computational domain boundaries

with the initial and boundary conditions:

$$\boldsymbol{u}(\bar{x}, 0) = \boldsymbol{u}_0 \qquad \boldsymbol{u}|_{\Gamma_1, \Gamma_4} = \boldsymbol{u}_b \qquad u_{\Gamma_2, \Gamma_3} = 0 \tag{3}$$

$$p_{\Gamma_2} = p_{in} \qquad p_{\Gamma_3} = p_{out} \tag{4}$$

where $\bar{x} = (x, y, z) \in \Omega$, $\boldsymbol{u} = (u, v, w)$ - velocity vector, $\boldsymbol{u}_b$ - velocity of the vessel walls and valve leaflets motion under deformation, $\rho = \rho(\bar{x}, t)$ - density, $p = p(\bar{x}, t)$ - pressure, $\sigma = \mu(\nabla \boldsymbol{u} + (\nabla \boldsymbol{u})^T)$ - viscous stress tensor, $\mu = \mu(\bar{x}, t)$ - fluid viscosity, $\boldsymbol{f} = \boldsymbol{f}(\bar{x}, t)$ - body forces vector, which is further used to determine form of the vessel and valve leaflets. Domain Ω is a vessel with boundary $\Gamma = \Gamma_1 \cup \Gamma_2 \cup \Gamma_3 \cup \Gamma_4$, where Γ_1 - blood vessel wall, Γ_2 and Γ_3 - inflow and outflow domains, Γ_4 - valve leaflets (see Fig. 2). As shown in [18], the problem (1) - (2) has a unique solution.

Density ρ and viscosity μ are defined by following relations [13]:

$$\mu = c(\mu_2 - \mu_1) + \mu_1 \tag{5}$$

$$\rho = c(\rho_2 - \rho_1) + \rho_1 \tag{6}$$

where ρ_1, μ_1 - fluid density and viscosity (plasma), ρ_2, μ_2 - admixture density and viscosity (formed elements), c - admixture concentration. Admixture concentration $c = c(\bar{x}, t)$, $c \in [0, 1]$ is determined as a solution of equation:

$$\frac{\partial c}{\partial t} + \boldsymbol{u} \cdot \nabla c = 0 \tag{7}$$

with initial conditions:

$$c(\bar{x}, 0) = c_0(\bar{x}), \bar{x} \in \Omega \tag{8}$$

and boundary conditions at the inflow boundary:

$$c(\bar{x}, t)|_{\Gamma_2} = c_s(\bar{x}, t) \tag{9}$$

where c_0, c_s are specified functions.

One of the issues determined for this kind of problem computational solutions is the lack of one component of velocity vector in the inflow-outflow areas. It can be solved by using the original equations (1) - (4) at the boundaries Γ_2, Γ_3 to determine the missing components of the velocity vector (see details [13]).

Motion of the vessel walls and valve leaflets is defined by the forces, which return them to the original position. Valve leaflets can be deformated much more, than vessel walls. To describe the forces, arising due to the valve deformation, the following formula is used:

$$F = \frac{\partial}{\partial s}(T\tau) + \frac{\partial^2}{\partial s^2}(E \cdot I \frac{\partial^2}{\partial s^2}X) \tag{10}$$

where $\bar{q} = (q, r, s) \in \Gamma_4$, $X(\bar{q})$ - function for describing the valve leaflets surface at the moment t, the coordinates q, r, s are chosen so that the surface X is presented by the large amount of parametric lines $s \to X(q^0, r^0, s)$, T - tension, that arises due to the stretching along s, E - Young's modulus, I - cross-sectional moment of inertia (see [7], [19]). Physically the formula above means, that the valve leaflets resist stretching, (it's related to the first term with T, which is dependent on stiffness coefficient k), and bending (it's related to the second term, where E and I are referred as a stiffness coefficient k_b). Formula (10) allows taking into account any changes of the valve shape.

To compute the forces, arising due to the deformation of the vessel, another formula is used, which allows taking into account only small shape changes:

$$F = k\|X - X_0\| \tag{11}$$

where $\bar{q} = (q, r, s, t) \in \Gamma_1$, $X(\bar{q}, t)$, $X_0(\bar{q}, 0)$ - functions for describing the surface of vessel walls at the moment t and at the initial time, k - stiffness coefficient.

Researches [6], [7] show, that in order to the interaction between vessel walls, valve leaflets and the fluid flow it is necessary to compute the field of external body forces f in the Navier-Stokes equation, based on the force F, and determine the current form $X(\bar{q}, t)$ of the vessel and the valve, based on the fluid field of velocities $u(\bar{x}, t)$. The following equations are used for this purpose:

$$\frac{\partial X}{\partial t}(\bar{q}, t) = \int_{\Omega} u(\bar{x}, t) \cdot \delta(x - X(\bar{q}, t)) \, dx \, dy \, dz \tag{12}$$

$$f(\bar{x}, t) = \int_{\Gamma} F(\bar{q}, t) \cdot \delta(x - X(\bar{q}, t)) \, dq \, dr \, ds \tag{13}$$

where $\bar{q} = (q, r, s) \in \Gamma$ - point at the vessel wall or valve leaflet, $X = X(\bar{q}, t)$ - function for describing the vessel and valve surfaces at the moment t, $F = F(\bar{q}, t)$ - the force of deformation resistance at given point, $u(\bar{x}, t)$ - fluid flow velocity vector, $f(\bar{x}, t)$ - body forces vector, δ - Dirac delta function.

Thus the model describing the motion of the viscous inhomogeneous incompressible fluid inside vessel with valve is built. This model enables to determine the fluid state and the surface form $\Gamma_1 \cup \Gamma_4$ independently of each other. The valve leaflets influence on the fluid is described by correlation (13) between the vector of body forces $\boldsymbol{f}(\bar{x}, t)$ from (1) and the force of deformation resistance $F = F(\bar{q}, t)$ from (10), (11).

3   Solution Method

As it was mentioned before, immersed boundary method is used in the paper [6]. This method is based on the fact, that in case of flowing over a body the fluid is effected by surface force and shear force if the body has no-slip boundary condition. The body surface is influenced by the same forces of opposite sign. It means that fluid flowing over the body can be modeled by a corresponding field of the external body forces [20].

According to the immersed boundary method, we determine the fluid flow in the parallelepiped $\tilde{\Omega}$, which contains Ω. $\tilde{\Omega}$ has no-slip boundary conditions. To compute of the fluid flow we use rectangular uniform staggered grid $\tilde{\Omega}_h$ with grid spacing h_x, h_y, h_z and staggered arrangement of cells, where the pressure, velocity divergence and concentration are computed at the center of cell, the velocity vector components and vector of external forces are computed at the boundaries of cell. To determine the deformation of the surface $\Gamma_1 \cup \Gamma_4$ we introduce additional area $\tilde{\Gamma}$ with Lagrangian coordinate system, which is related to the vessel walls and valve leaflets. In the $\tilde{\Gamma}$ we construct a new grid $\tilde{\Gamma}_h$, with cells corresponding to the points at the $\Gamma_1 \cup \Gamma_2$. Solution algorithm consists of several steps: at the grid $\tilde{\Gamma}_h$ the problem (1)-(4) is solved; then the convection equation (7) is solved, i.e. the concentration of admixture is determined in the solution domain and the density and viscosity are recalculated. Then formulas (10), (11) and (12), (13) are used to determine the position of leaflets and the vessel form.

Differential equation (1), (9) is solved by the finite difference method. To solve (1), (4) splitting schemes due to physical factors are used [21]:

$$\frac{u^* - u^n}{\triangle t} = -(u^n \cdot \nabla)u^* - \frac{1}{\rho}\nabla\sigma + f^n \tag{14}$$

$$\rho\triangle p^{n+1} - \nabla\rho \cdot p^{n+1} = \frac{\rho^2\nabla u^*}{\triangle t} \tag{15}$$

$$\frac{u^{n+1} - u^*}{\triangle t} = -\frac{1}{\rho}\triangle p^{n+1} \tag{16}$$

Numerical implementation of this scheme consists of three stages. At the beginning the intermediate field u^* is computed using the known values of velocity from the previous time step. Thus equation (14) is solved by the method of stabilizing corrections [22]. Then a new pressure field is determined via the computational solution of (15) using biconjugate gradient method. At the last stage a final velocity vector field is calculated according to the formula (16).

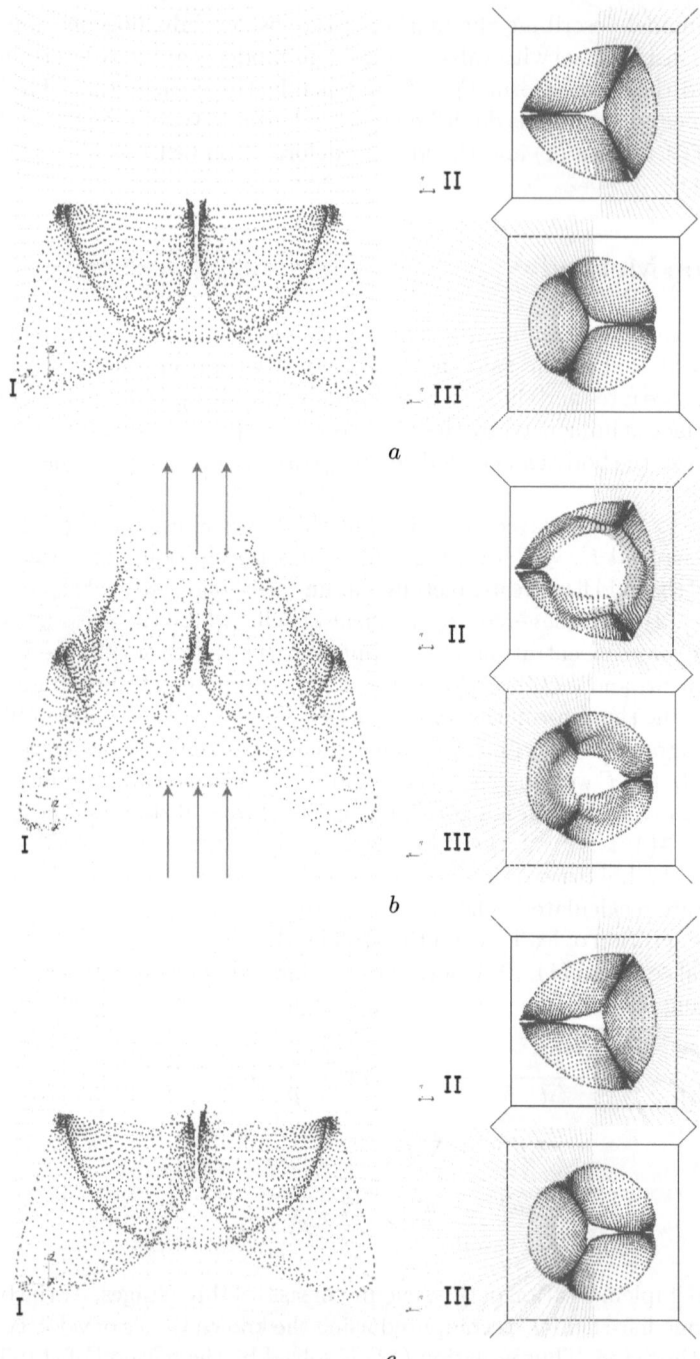

Fig. 3. Dynamics of the Valve leaflets. Current leaflet shape is indicated by points, arrows indicate flow direction. Side view (I), front view (II) and rear view (III). $k_s = 5 \cdot 10^3$, $k_b = 5 \cdot 10^3$, $\rho_1 = \rho_2 = 1$, $\mu_1 = \mu_2 = 1 \cdot 10^{-2}$; a) $t = 0$, b) $t = 0.7$, c) $t = 1.5$

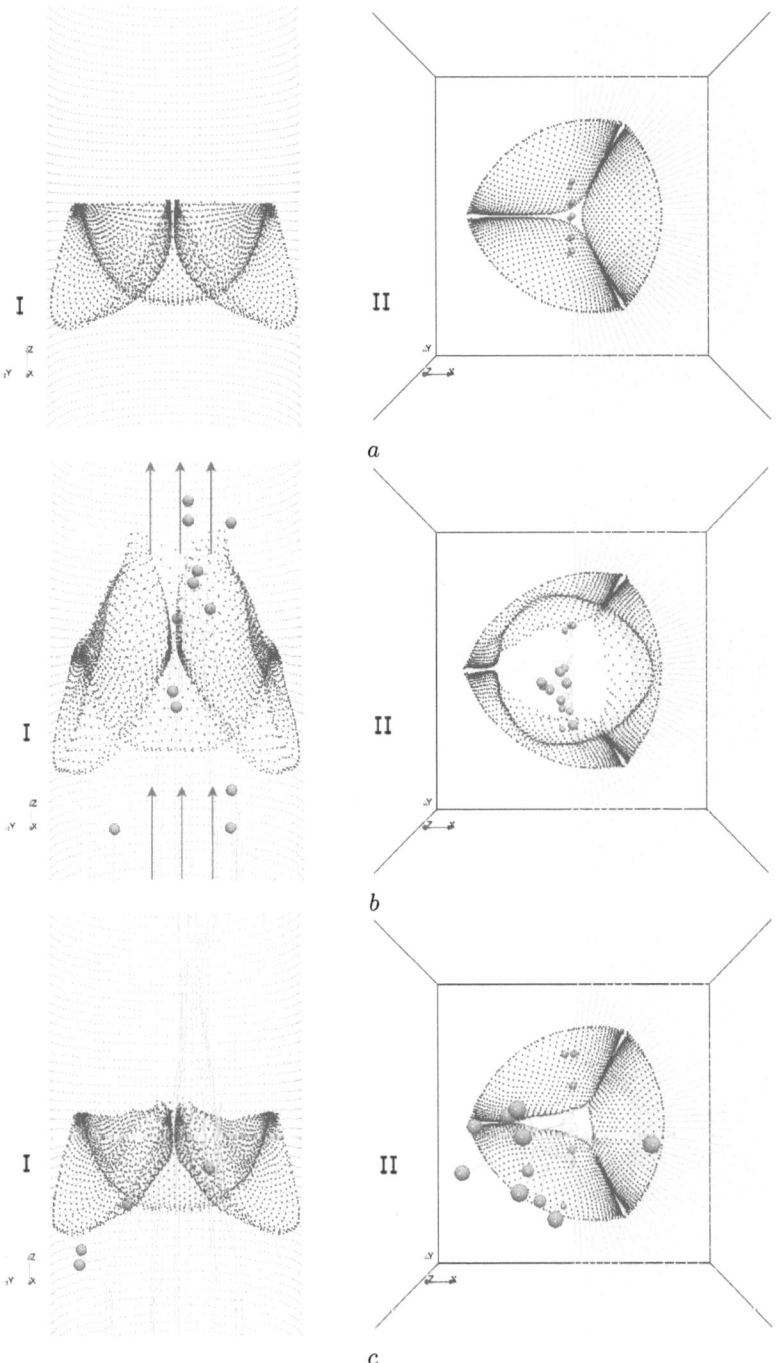

Fig. 4. Tracks of particles inside the valve. Flow direction is indicated by arrows. Calculation parameters are the same as in the Fig. 3 Side view (I) and front view (II). $k_s = 5 \cdot 10^3$, $k_b = 5 \cdot 10^3$, $\rho_1 = \rho_2 = 1$, $\mu_1 = \mu_2 = 1 \cdot 10^{-2}$; a) $t = 0$, b) $t = 0.7$, c) $t = 1.5$

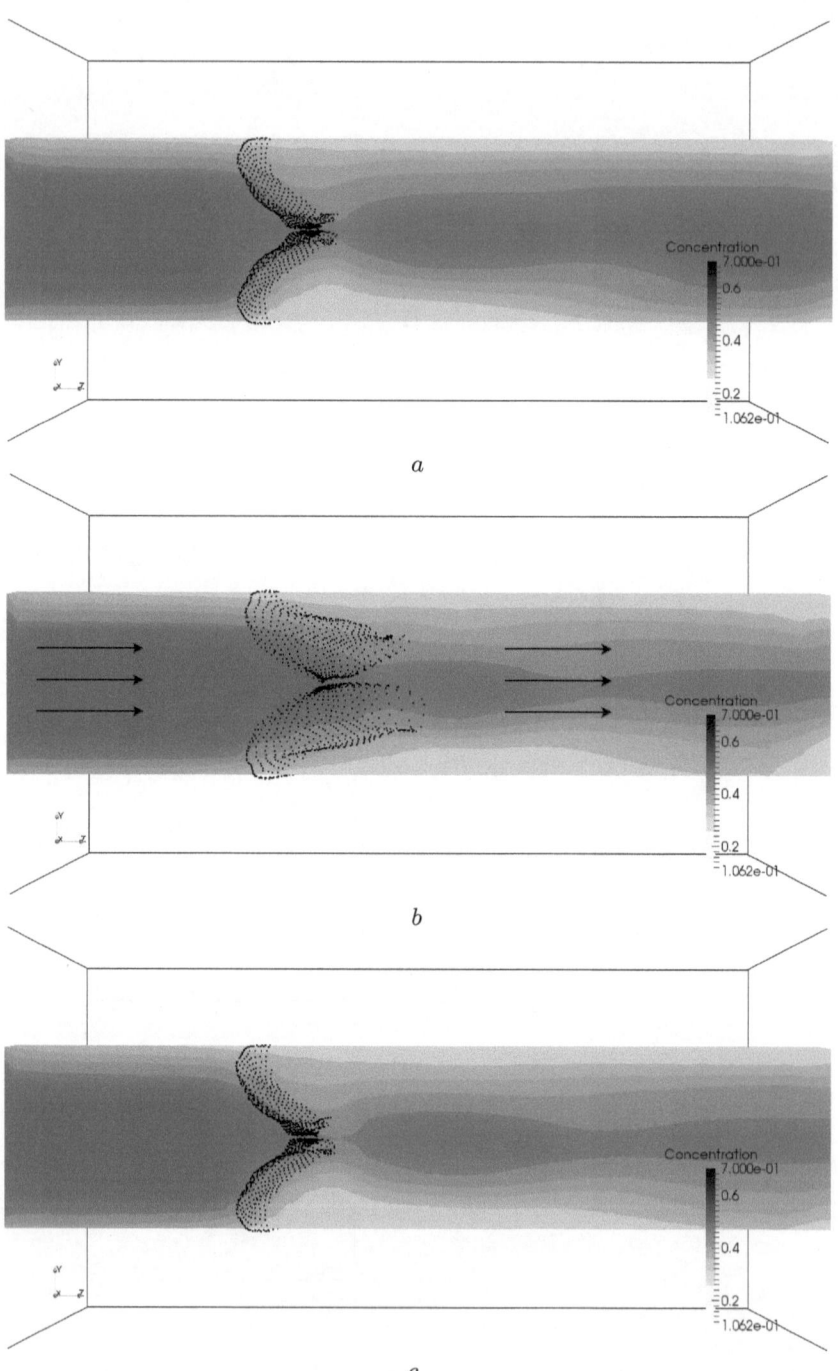

Fig. 5. Valve leaflets motion in a vessel with variable viscosity and density. A constant admixture flow $c_s|_{\Gamma_2} = 0.45$ at the inflow, admixture concentration at the initial time $c_0 = 0.45$, $\rho_1 = 1$, $\rho_2 = 1.2$, $\mu_1 = 1 \cdot 10^2$, $\mu_2 = 1.2 \cdot 10^2$; a) $t = 4$, b) $t = 5$, c) $t = 6$

As soon as fluid flow parameters are determined it is necessary to calculate new values of density and velocity. To do that a new time step for the convection equation (7) must be done using the obtained values of velocity components, and the density and viscosity are recalculated by the formulas (5), (6).

Next it is necessary to determine the deformation of vessel walls and valve leaflets being influenced by of fluid flow, and also the distribution of body forces f in the fluid motion equation based on this deformation. It is possible to calculate the deformation of vessel walls and valve leaflets under this particular fluid pressure and the resistance forces by using the equations (12) - (13), which are numerically integrated by using any of quadrature formulas, and equations (10) - (11). Afterwards, the body forces f are recalculated, and it is possible to move to the next time step.

4   Results

Some results of methodical calculations for the cases with constant and variable density and viscosity, which are aimed to demonstrate the described method validation and the possibility to get the patterns of leaflet deformation and admixture distribution inside the valve. All calculations were performed in dimensionless variables. A circular cylinder with length $l = 1$, radius $r = 0.11$ and wall stiffness $k = 1 \cdot 10^3$ was used as a vessel with a valve, the domain $\tilde{\Omega}$ had spatial parameters $1.0 \times 0.5 \times 0.5$, spatial steps $h_x = h_y = h_k = 0.01$, time step $\triangle t = 0.01$.

Fig. 3 and Fig. 4 show tricuspid valve dynamics effected by the pressure of fluid with constant density and velocity. The pressure differential $p_{in} - p_{out}$ changes periodically from 0 to 6. Coefficient of stretching resistance $k_b = 5 \cdot 10^3$ and coefficient of bending resistance $k_b = 5 \cdot 10^3$ are specified for the valve leaflets.

As can be seen in the Fig. 3 and Fig. 4, the valve opens when pressure differential is increased, and then reverts to the original state when pressure is balanced.

The Fig. 5 shows the motion dynamics of the tricuspid valve under the pressure of fluid with variable viscosity and density. Pressure differential $p_{in} - p_{out}$ changes cyclically from 0 to 6. Coefficient of stretching resistance $k_s = 8 \cdot 10^3$ and coefficient of bending resistance $k_s = 6 \cdot 10^3$ are specified for the valve leaflets. Constant admixture flow with concentration $c_s = 0.45$ is set at Γ_2.

Fig. 5 shows, that initial uniform admixture distribution is interrupted by the valve leaflets motion. Eventually oscillatory mode of admixture motion can be recognized that corresponds to valve operating cycle. Moreover, the Fig. 5 shows that admixture distribution over a cross section being parallel to axis Oy is not symmetric because the valve leaflets are not symmetric about the axis Oy as well.

5   Conclusion

Constructed model of blood flow with variable viscosity and density allows to get the patterns of leaflet deformation and admixture distribution effected by inhomogeneous fluid flow.

Acknowledgments. This research is performed as part of the government contract 1.630.1.2014/K.

References

1. Yoganathan, A.P., He, Z.M., Jones, S.C.: Fluid mechanics of heart valves. Annu. Rev. Biomed. Eng. **6**, 331–362 (2004)
2. Yacoub, N., Takkenberg, J.: Will heart valve tissue engineering change the world? Nat. Clin. Prac. Cardiovas. Med. **2**, 60–61 (2005)
3. Taylor, C.A., Hughes, T.J.R., Zarins, C.K.: Finite Element Modeling of Blood Flow in Arteries. Computer Methods in Applied Mechanics and Engineering **158**, 155–196 (1998)
4. Zhang Y, Bajaj C.: Finite element meshing for cardiac analysis. ICES Technical Report, pp. 4–26 (2004)
5. Black, M.M., Howard, I.C., Huang, X., Patterson, E.A.: A three-dimensional analysis of a bioprosthetic heart valve. J. Biomech. **24**(9), 793–801 (1991)
6. Peskin, C.S.: Numerical Analysis of Blood Flow in the Heart. JCP **25**, 220–252 (1977)
7. Boyce E.G.: Immersed boundary model of aortic heart valve dynamics with physiological driving and loading conditions. International Journal for Numerical Methods in Biomedical Engineering 1–29 (2011)
8. Ma, X., Gao, H., Boyce, E.G., Berry, C., Luo, X.: Image-baseduidstructure interactionmodel of the human mitral valve. Computers & Fluids **71**, 417–425 (2013)
9. Pilhwa L., Boyce E.G., Peskin C.S.: The immersed boundary method for advection-electrodiffusion with implicit time stepping and local mesh refinement. Comput. Phys. **229**(13) (2010)
10. Fai, T.G., Boyce, E.G., Mori, Y., Peskin, C.S.: Immersed boundary method for variable viscosity and variable density problems using fast constant-coefficient linear solvers I: numerical method and results. SIAM Journal on Scientific Computing **35**(5), B1132–B1161 (2013)
11. Jian, D., Robert, D.G., Aaron, L.F.: An immersed boundary method for two fluid mixtures. Journal of Computational Physics **262**, 231–243 (2014)
12. Lee, P., Boyce, E.G., Peskin, C.S.: The immersed boundary method for advection-electrodiffusion with implicit time stepping and local mesh refinement. Journal of Computational Physics **229**, 5208–5227 (2010)
13. Gummel, E.E., Milosevic, H., Ragulin, V.V., Zakharov, Y.N., Zimin, A.I.: Motion of viscous inhomogeneous incompressible fluid of variable viscosity. Zbornik radova konferencije MIT 2013, Beograd, pp. 267–274 (2014)
14. Geidarov, N.A., Zakharov, Y.N., Shokin, Y.I.: Solution of the problem of viscous fluid flow with a given pressure differential. Russian Journal of Numerical Analysis and Mathematical Modeling **26**(1), 39–48 (2011)
15. Milosevic, H., Gaydarov, N.A., Zakharov, Y.N.: Model of incompressible viscous fluid flow driven by pressure difference in a given channel. International Journal of Heat and Mass Transfer **62**, 242–246 (2013). July 2013. ISSN: 0017–9310
16. Dolgov, D.A., Zakharov, Y.N.: Modeling of viscous inhomogeneous fluid flow in large blood vessels. Vestnik Kemerovo State University **2**(62) T.1, pp. 30–35 (2015) (in Russian)
17. Caro, C.G., Pedley, J., Schroter, R.C., Seed, W.A.: The Mechanics of the Circulation. Moscow: Mir, 624 (1981)
18. Ragulin, V.V.: To the problem of flow viscous fluid through the limited area under given pressure differential. Dynamic of Continuum: Novosibirsk **27**, 78–92 (1976)

19. Peskin, C.S.: The immersed boundary method. Acta Numerica **11**, 479–517 (2002)
20. Goldstein, D., Handler, R., Sirovich, L.: Modeling a no-slip flow boundary with an external force field. Journal of Computational Physics **105**, 354–366 (1993)
21. Belotserkovskii, O.M.: Numerical modeling in mechanics of continuum. Moscow: Science, 520 (1984) (in Russian)
22. Yanenko, N.N.: Method of fractional steps for solving multidimensional problems of mathematical physics. Novosibirsk: Science, 197 (1967) (in Russian)

Numerical Model of Plasma-Chemical Etching of Silicon in CF_4/H_2 Plasma

Aleksey Gorobchuk[1,2($\boxtimes$)]

[1] Institute of Computational Technologies SB RAS, Av. Lavrentiev 6,
630090 Novosibirsk, Russia
[2] Novosibirsk State University, Pirogova Str. 2, 630090 Novosibirsk, Russia
alg@eml.ru
http://www.ict.nsc.ru

Abstract. The 2D mathematical model of plasma-chemical etching process, where the gas flow of the mixture was described by the equations of multicomponent physical-chemical hydrodynamics, was presented. The silicon etching in CF_4/H_2 gas mixture was studied. The chemical kinetic model contained 28 gas-phase reactions of dissociation and recombination processes and 6 heterogeneous reactions on the wafer, which included the products - F, F_2, CF_2, CF_3, CF_4, C_2F_6, H, H_2, HF, CHF_3, CH_2F_2. The concentrations of chemical components were calculated from the system of conservation equations included the mentioned gas-phase reactions. The governing equations were numerically solved by iterative finite difference splitting-up method. It is shown that the CF_4/H_2 system is characterized by lower fluorine concentrations and higher CF_2, CF_3 coverage of silicon surface compared to the CF_4/O_2 system.

Keywords: Mathematical modeling · Numerical methods · Plasma-chemical etching technology · Multicomponent gas mixtures

1 Introduction

The processing of thin films by fluorine atoms in plasma-chemical reactors is widely used in microelectronic device production. The active particles are formed in the RF-discharge zone by the dissociation of gas molecules containing such atoms. Usually they are pure gases CF_4, SF_6 or binary gas mixtures with O_2, H_2 and etc. Due to the complex multichannel nature of fluorine formation the probable mechanisms of gas-phase chemical reactions in glow discharge are insufficient investigated. Surface phenomena at the RF-electrodes and wafer surface are even less understood. The defining set of chemical reactions in plasma is usually chosen using the experimental results. The concentrations of active particles strongly depend on the choice of chemical kinetic model and the electron density distribution in RF-discharge. To provide a good optimization of the etching process is quite essential to compare the probable kinetic models of fluorine formation in such chemical systems. It may be fulfilled on the base of mathematical

© Springer International Publishing Switzerland 2015
N. Danaev et al. (Eds.): CITech 2015, CCIS 549, pp. 44–52, 2015.
DOI: 10.1007/978-3-319-25058-8_5

modeling. Some results obtained for chemical kinetic model of silicon etching in CF_4/O_2 glow discharge plasma have shown that to obtain adequate results it is necessary to use a detail plasma-chemical kinetics with precise description of heat and mass transfer [1]. Thereat a simulation has to incorporate calculating the hydrodynamical and molecular transport processes in the etching chamber.

In the paper one of probable mechanisms of gas-phase chemical reactions in CF_4/H_2 mixture and their influence on the etching process are studied.

2   Mathematical Model Formulation

The calculations were carried with using 2D mathematical model of plasma-chemical etching reactor [2] in which a special attention gives to the multicomponent chemical kinetics of gas-phase reactions.

In CF_4/H_2 mixture the basic set of chemical reactions corresponding to reactions in pure CF_4 was derived. Further the chemical reaction set was added by possible reactions of CF_4 with H_2. Consequently the chemical kinetic model contains 28 gas-phase reactions of dissociation and recombination processes, which include F, CF_2, CF_3, CF_4, C_2F_6, F_2, H, H_2, HF, CHF_3, CH_2F_2 [3]:

$$CF_4 + e \longrightarrow CF_3 + F + e, \tag{1}$$

$$CF_4 + e \longrightarrow CF_2 + 2F + e, \tag{2}$$

$$H_2 + e \longrightarrow H + H + e, \tag{3}$$

$$CHF_3 + e \longrightarrow CF_3 + H + e, \tag{4}$$

$$CF_3 + CF_3 + M \longrightarrow C_2F_6 + M, \tag{5}$$

$$F + CF_3 + M \longrightarrow CF_4 + M, \tag{6}$$

$$F + CF_2 + M \longrightarrow CF_3 + M, \tag{7}$$

$$F + CHF_3 \longrightarrow CF_3 + HF, \tag{8}$$

$$F + CHF_3 \longrightarrow CF_4 + H, \tag{9}$$

$$F + H_2 \longrightarrow HF + H, \tag{10}$$

$$F + H + M \longrightarrow HF + M, \tag{11}$$

$$F + F + M \longrightarrow F_2 + M, \tag{12}$$

$$F_2 + M \longrightarrow F + F + M, \tag{13}$$

$$H + CF_4 \longrightarrow CF_3 + HF, \tag{14}$$

$$H + CF_3 \longrightarrow CF_2 + HF, \tag{15}$$

$$H + CHF_3 \longrightarrow CF_3 + H_2, \tag{16}$$

$$H + CF_4 \longrightarrow CHF_3 + F, \tag{17}$$

$$H + CF_3 + M \longrightarrow CHF_3 + M, \tag{18}$$

$$H + HF \longrightarrow F + H_2, \tag{19}$$

$$H + H + M \longrightarrow H_2 + M, \tag{20}$$

$$H_2 + CF_3 \longrightarrow CHF_3 + H, \tag{21}$$

$$H_2 + CF_2 + M \longrightarrow CH_2F_2 + M, \tag{22}$$

$$HF + CF_3 \longrightarrow CF_4 + H, \tag{23}$$

$$HF + CF_2 \longrightarrow CF_3 + H, \tag{24}$$

$$HF + CF_2 + M \longrightarrow CHF_3 + M, \tag{25}$$

$$HF + CF_3 \longrightarrow CHF_3 + F, \tag{26}$$

$$CHF_3 + M \longrightarrow CF_2 + HF + M, \tag{27}$$

$$CH_2F_2 + M \longrightarrow CF_2 + H_2 + M. \tag{28}$$

The reactions (1)-(4) describe the dissociation of CF_4, H_2 and CHF_3 molecules by electron-impact generating chemical active atoms of fluorine and hydrogen; the reactions (5)-(28) represent the volume recombination of reactive atoms and radicals with third body M.

The chemical kinetics of heterogeneous reactions was presented by processes of adsorption of CF_2, CF_3 at wafer surface. In all 6 heterogeneous reactions are considered:

$$CF_3 \xrightarrow{k_{s1}} CF_3(s), \tag{29}$$

$$CF_2 \xrightarrow{k_{s2}} CF_2(s), \tag{30}$$

$$F + CF_2(s) \xrightarrow{k_{s3}} CF_3, \tag{31}$$

$$F + CF_3(s) \xrightarrow{k_{s4}} CF_4, \tag{32}$$

$$CF_3 + CF_3(s) \xrightarrow{k_{s5}} C_2F_6, \tag{33}$$

$$4F + Si \xrightarrow{k_s} SiF_4. \tag{34}$$

Here reactions (29)-(33) are heterogeneous reactions of adsorption-desorption of CF_2 and CF_3 radicals on the silicon surface; $k_{s1} - k_{s5}$ are rate constants of heterogeneous reactions; the reaction (34) is the reaction of spontaneous silicon etching; is the etching rate constant. The designation (s) is used for radicals adsorbed on the wafer surface.

According to the selected model of multicomponent chemical kinetics the distribution of concentration for each component was calculated from the system of interconnected equations of convective-diffusion transfer:

$$\mathbf{v} \cdot \nabla C_i = \nabla \cdot (D_{i-m} C_t (\nabla x_i)) + G_i(C_i, C_j)$$

where $i, j = F$, CF_2, CF_3, CF_4, C_2F_6, F_2, H, H_2, HF, CHF_3, CH_2F_2; C_i, x_i are the mole fraction and mole concentration of particles i; C_t is the total gas concentration; D_{i-m} is the multicomponent diffusion coefficient of particles i; G_i is the generation rate of particles i by gas-phase reactions. The gas flow (vector $\mathbf{v}$) was described by the equations of multicomponent physical-chemical hydrodynamics.

The right part of the system of convective-diffusion equations includes the base set of crucial gas-phase reactions (1)-(28) which define a complex interconnection between the particle generation rates. The source term carries in to equations the power nonlinearity concerning particle concentrations. Moreover the spatial distribution of initial electron density in discharge plasma defines the generation rates of active particles [4]. Depending on the pressure and gas medium that is under consideration, the dominant electron loss mechanism can be diffusion, recombination or attachment. It was assumed that the electron density distribution corresponded to a diffusion-controlled approach. In the parametric calculations the simplified model of radio frequency discharge between two plane electrodes was used.

According to the set of heterogeneous reactions the passivation of silicon surface by adsorbed particles takes place. The different parts of silicon surface are covered by various adsorbed atoms and radicals. The competing adsorption of radicals CF_2, CF_3 on silicon overlap the access of fluorine atoms to the wafer and decrease spontaneous etching silicon. The fractions of silicon surface covered by radicals CF_2, CF_3 designate as ϑ_{CF_2} and ϑ_{CF_3} accordingly. The balances of mass flows for CF_2 and CF_3 components on silicon surface at equilibrium give the following relations for unknown parameters ϑ_{CF_2} and ϑ_{CF_3} [4]:

$$k_{s2}x_{CF_2}/k_{s3}x_F = \vartheta_{CF_2}/(1 - \vartheta_{CF_2} - \vartheta_{CF_3}),$$

$$k_{s1}x_{CF_3}/(k_{s4}x_F + k_{s5}x_{CF_3}) = \vartheta_{CF_3}/(1 - \vartheta_{CF_2} - \vartheta_{CF_3}).$$

The solution of this equations are the next formulas:

$$\vartheta_{CF_2} = k_{s2}x_{CF_2}/(k_{s2}x_{CF_2} + k_{s3}x_F + \Delta_{CF_2}),$$

$$\vartheta_{CF_3} = k_{s1}x_{CF_3}/(k_{s1}x_{CF_3} + k_{s4}x_F + k_{s5}x_{CF_3} + \Delta_{CF_3}),$$

where

$$\Delta_{CF_2} = k_{s1}x_{CF_3}k_{s3}x_F/(k_{s4}x_F + k_{s5}x_{CF_3}),$$

$$\Delta_{CF_3} = k_{s2}x_{CF_2}(k_{s4}x_F + k_{s5}x_{CF_3})/k_{s3}x_F.$$

In the present formulas the unknown characteristics of adsorption layers have a simple presentation as the function of rate constants of chemical reactions (29)-(33).

The heterogeneous and silicon etching reactions entered into a boundary conditions at the wafer. The latter were written as a balance of mass flows for each component. The presented fractions of silicon surface covered by CF_2, CF_3 are very important parameters because they are used in the boundary conditions and the etching rate. For example, the local spontaneous etching rate in Å/min was defined by the formula:

$$v_s = 1.81 \cdot 10^{10} \left(1 - \vartheta_{CF_2} - \vartheta_{CF_3}\right) k_s C_F,$$

where k_s is etching rate constant, cm/s; C_F is mole concentration of fluorine, Mol/cm^3. Because of adsorption CF_2, CF_3 on the wafer the etching rate includes a complicated nonlinear dependence on concentrations of components F, CF_2, CF_3, which shows competitive mechanism of particle interaction with silicon surface.

3   Numerical Method

The solution of problem was carried out by the numerical finite-difference method briefly presented in [1]. The calculating domain was covered by uniform grid Ω_h with 76×46 mesh points. A presence of two-order elliptic operators in all equations of the mathematical model allow us to approximate each equation by implicit iterative finite-difference splitting-up scheme with stabilizing correction. The scheme in general form looks as follows:

$$\frac{\phi^{k+1/2} - \phi^k}{\tau} = L_\xi^\phi \phi^{k+1/2} + L_\zeta^\phi \phi^k + F(\phi^k),$$

$$\frac{\phi^{k+1} - \phi^{k+1/2}}{\tau} = L_\zeta^\phi (\phi^{k+1} - \phi^k),$$

where ϕ^k is the mesh function of solution at the time iteration k; τ is iterative parameter. An approximation order is $O(\tau + h_1^2 + h_2^2)$, where h_1, h_2 are mesh widths along ξ and ζ coordinates.

The hydrodynamic equations of Navier-Stokes were reduced by using variables "stream function - vorticity" (ψ, ω). The directed finite difference of second order was applied to approximate the vorticity at walls of reactor chamber, for example, in mesh point (n, w):

$$\omega_{n,w}^k = \frac{2}{\xi_n} \frac{(\psi_{n,w-2}^k - \tilde{r}_{2,w-2}^3 \psi_{n,w-1}^k - (1 - \tilde{r}_{2,w-2}^3)\psi_{n,w}^k)}{h_{2,w-2}h_{2,w-1}\tilde{r}_{2,w-2}^2},$$

where $\tilde{r}_{2,w-2} = 1 + h_{2,w-2}/h_{2,w-1}$; $h_{2,w}$ is mesh width along ζ coordinate in point w. The stream function was find for each iteration of vorticity.

The concentrations of reagents x_i were calculated from the convective-diffusion equations using the computed velocity. To approximate the diffusion fluxes of reagents Q_d at walls of reactor chamber the directed finite difference of second order was used, for example, in mesh point (w, j) normal to the direction ξ:

$$Q_{d_{w,j}}^k \approx -\tilde{d}_i \tilde{c}_t \frac{(1 + 2\tilde{e}_{1,w-1})x_{i_{w,j}}^k - (1 + \tilde{e}_{1,w-1})^2 x_{i_{w-1,j}}^k + \tilde{r}_{1,w-1}^2 x_{i_{w-2,j}}^k}{h_{1,w-1}(1 + \tilde{e}_{1,w-1})},$$

where $\tilde{e}_{1,w-1} = h_{1,w-1}/h_{1,w-2}$; $\tilde{d}_i, \tilde{c}_t$ are normalized coefficient of multicomponent diffusion of reagent i and total gas concentration respectively. The etching rate of wafer was then found for known concentrations of reagents.

A solution of an original steady state problem was derived by relaxation method. The iterative process was ended with achieving of relative error $\varepsilon_\phi = 10^{-10} - 10^{-8}$ in uniform norm:

$$\max_{\Omega_h} \left| \frac{\phi^{k+1} - \phi^k}{\tau \phi^{k+1}} \right| < \varepsilon_\phi.$$

The created adequate numerical model of reactor process allows to investigate subtle physical effects of plasma-chemical etching.

4   Results and Discussion

The effect of multicomponent plasma kinetics on the production and mass transfer of active particles was studied on example of radial flow plasma-chemical etching reactor. The construction dimensions are used as in [2]. The gas flow direction to the center of reactor was examined. The calculations have been done for gas flow rate under normal conditions $Q = 200$ cm^3/min. The pressure in etching chamber of reactor was equal to $p = 0.5$ torr. The temperature of reactor walls and wafer were $T_w = T_s = 300$ K. The average electron density was assumed equal to $\overline{n}_e = 6 \times 10^9$ cm^{-3}. The H_2 percentage fraction in CF_4/H_2 feed gas mixture varied in the range 0 - 90 %.

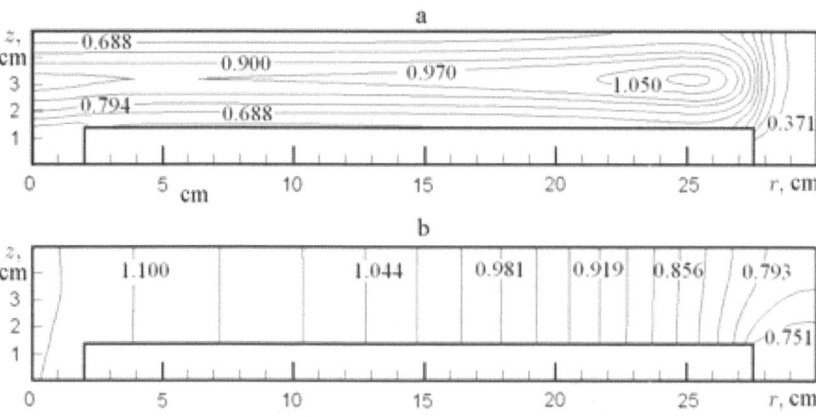

Fig. 1. The distributions of fluorine (a) $C_F \times 10^{-10}$ Mol/cm^3 and hydrofluorine (b) $C_{HF} \times 10^{-8}$ Mol/cm^3 concentrations in the radial flow reactor. Processing regime: $p = 0.5$ torr, $Q = 200$ cm^3/min, $T_s = 300$ K, 25 % fraction of H_2 in CF_4/H_2.

The distributions of concentrations F and HF are shown on Fig. 1, a and Fig. 1, b correspondingly at 25 % fraction H_2 in CF_4/H_2. The main part of fluorine atoms obtained from the dissociation of tetrafluoromethane are consumed in the reactions with hydrogen atoms to form an abundant component HF. As a result the fluorine concentration monotonous decreases along the flow direction whereas the concentration HF rises and reaches a maximum at the outlet of reactor.

With increase of H_2 percentage in the feed gas mixture over the range 0-40 % the fraction of silicon wafer occupied by radicals CF_2 rises proportionally with the concentration of H_2 (see Fig. 2) and riches the value 0.98797 at 40 % H_2. The fraction of silicon surface covered by radicals CF_3 not exceeds the value 0.01563, which is reached at 25 % H_2 and in further calculations may not take into account. Starting with 40 % addition H_2 the all silicon surface becomes passive because of the intensive adsorption of radicals CF_2, CF_3 for which $\vartheta_{CF_2} + \vartheta_{CF_3} = 0.99061$. On the contrary in CF_4/O_2 system the fractions ϑ_{CF_2} and ϑ_{CF_3} beginning from 5 % O_2

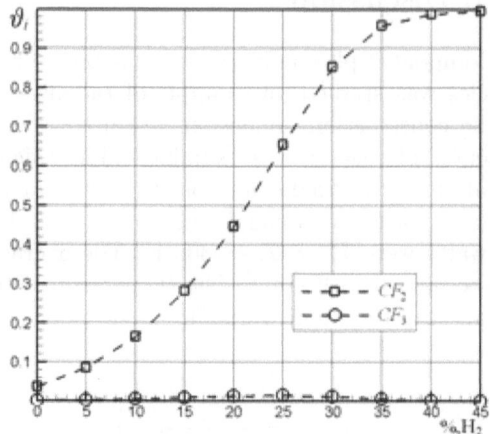

Fig. 2. The fractions of silicon surface covered by adsorbed radicals CF_2 and CF_3 versus inlet H_2 addition in CF_4/H_2 mixture.

are less than 0.01 because of general depletion of gas mixture by fluorine-containing radicals CF_x intensively interacting with atomic oxygen.

The calculation concentrations of components at the wafer are presented on Fig. 3 as function of percentage fraction of H_2 in CF_4/H_2. The large part of fluorine is consumed in reactions with atomic hydrogen. With increase of H_2 percentage in the feed gas composition the fluorine concentration rapidly decreases because much of it is used to form particles HF. The concentration of HF linearly rises with the increase of H_2 and has a maximum at 50 %. The fluorine concentration reaches a very low value at 50 % H_2 and than practically disappears. The second abundant concentration after CF_4 is HF which is reaches a maximum at 45-50 % H_2 in the mixture. In the range of addition of H_2 0-30 % the concentration CF_2 rises very slowly but starting with 30 % sharply increases and has a maximum at 50 % H_2. In the range of H_2 addition 10-50 % the values of F, CF_3 concentrations are over order lower then the concentrations of other components CF_4, CF_2, H_2 and HF. The fluorine concentration monotonically falls down as H_2 content rises. The concentration of CF_3 weakly rises at 0-25 % H_2 and then falls down too. As a result of chemical reactions the main stable products HF, CHF_3, CH_2F_2 are formed. The concentrations of CHF_3, CH_2F_2 are essentially smaller then the most abundant concentrations (over two order).

In CF_4/O_2 system the oxygen atoms replacement fluorine atoms in fluorine-containing radicals CF_x which set free the additional F. Thereupon the fluorine concentration in CF_4/H_2 system is considerably lower because of its additional consumption in the reaction with hydrogen to form HF.

The fluorine component is weakly consumed in the etching reaction owing to the passivation of silicon surface by the adsorbed radicals CF_2, CF_3. Moreover the fluorine concentration decreases in general. The adsorption of CF_2 on the

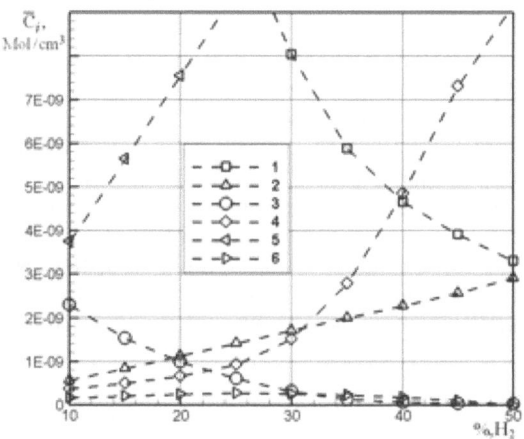

Fig. 3. Average concentrations of chemical components near the wafer as a function of percentage fraction of H_2 in CF_4/H_2 mixture. The designations: 1 - CF_4, 2 - H_2, 3 - F, 4 - CF_2, 5 - HF, 6 - CF_3. The concentrations of F and CF_3 components are increased on the order above. Processing regime: see Fig. 1.

surface results in to the reduction of etching rate when the fluorine concentration is nonzero. The etching process is completely stop at 35 % H_2. The main channel of reduction of silicon etching in CF_4/H_2 connects with two processes − the fluorine depletion and the surface passivation by radicals CF_2. Thus the hydrogen addition up to 30 - 40 % allows to decrease the etching rate and it is an effective factor for controlling the processing regime.

5   Conclusion

The simulation of silicon etching process in CF_4/H_2 plasma allows to conclude the following results.

The CF_4/H_2 system is characterized by lower fluorine concentration and higher coverage of silicon surface by CF_2, CF_3 compared to the CF_4/O_2 system. The most substantial components after CF_4 are HF and CF_2 which reach a maxima at 50 % addition of H_2. The most part of fluorine goes on formation of component HF. With increase of H_2 percentage in the feed gas mixture the fraction of silicon wafer occupied by radicals CF_2 rapidly rises and at 40 % H_2 comes to 99 %. The fraction of silicon wafer covered by CF_3 not exceeds 1.5 % in all range of parameters. The addition of H_2 in the limits up to 35 % allows to completely stop the etching process and is an effective factor to control the processing regime.

This research was supported by the Russian Fund of Basic Research (grant No.14-01-00274) and by the grant of the President of Russian Federation for state supporting of scientific school (grant No.5006.2014.9).

References

1. Grigoryev, Y.N., Gorobchuk, A.G.: Micro electronic and mechanical systems: numerical simulation of plasma-chemical processing semiconductors. In: Takahata, K. (ed.) In-Tech Education and Publishing (2009)
2. Grigoryev, Y.N., Gorobchuk, A.G.: Specific Features of Intensification of Silicon Etching in CF_4/O_2 Plasma. Russian Microelectronics **36**, 321–332 (2007)
3. Grigoryev, Y.N., Gorobchuk, A.G.: Simulation of the polymerization process on a silicon surface under plasma-chemical etching in CF_4/H_2. Journal of Surface Investigation X-ray, Synchrotron and Neutron Techniques **9**, 184–189 (2015)
4. Grigoryev, Y.N., Gorobchuk, A.G.: Effect of HF Discharge Structure on Etch Nonuniformity in Plasma-Chemical Reactor. Russian Microelectronics **43**, 34–41 (2014)

Simulation of Transonic Airfoil Flow Using a Zonal RANS-LES Method

Alibek Issakhov[1]([✉]), Benedikt Roidl[2], Matthias Meinke[2], and Wolfgang Schröder[2]

[1] Al-Farabi Kazakh National University, Al-Farabi ave. 71, 050040 Almaty, Kazakhstan
alibek.issakhov@gmail.com
[2] Institute of Aerodynamics, RWTH Aachen University, Wüllnerstraße 5a, Aachen, Germany

Abstract. This paper presents a method for a synthetic turbulence generation (STG) to be used in a segregated hybrid Reynolds-averaged Navier-Stokes (RANS)-Large-Eddy Simulation (LES) approach. The present method separates the LES inflow plane into three sections where a local velocity signal is decomposed from the turbulent flow properties of the upstream RANS solution. Depending on the wall-normal position in the boundary layer, the local flow Reynolds and Mach number specific time, length and velocity scales with different vorticity contents are imposed on the LES inflow plane. The STG method is assessed by comparing the resulting skin-friction, velocity and Reynolds-stress distributions of zonal RANS-LES simulations of flat plate boundary layers with available pure LES, DNS, and experimental data. It is shown that for the presented flow cases a satisfying agreement within a short RANS-to-LES transition of two boundary-layer thicknesses is obtained. The method is further used for the simulation of a shock-boundary-layer interaction around an airfoil at transonic flow conditions, where the separated flow region are analyzed by an embedded LES and the remaining flow is determined by a RANS solution.

Keywords: Zonal RANS/LES · Synthetic turbulence · Boundary layer

1 Introduction

CFD simulations at high Reynolds numbers for technical applications are nowadays mainly based on solutions of the Reynolds averaged Navier-Stokes (RANS) equations. The main reason are that they are simple to apply and computationally more efficient than other turbulence modelling approaches such as LES.It is known, however, that in many flow problems the condition of a turbulent equilibrium is not satisfied, i.e., when strong pressure gradients or flow separation occurs, which reduces the prediction accuracy of the results obtained by one- and two-equation turbulence models used to close the RANS equations [13,15].

Alternatives to RANS solutions are direct numerical and large-eddy simulation (DNS and LES). The limits of todays available computer resources, however, still prevent these methods to become standard simulation tools for high

© Springer International Publishing Switzerland 2015
N. Danaev et al. (Eds.): CITech 2015, CCIS 549, pp. 53–65, 2015.
DOI: 10.1007/978-3-319-25058-8_6

Reynolds number flows. In many technical flow problems complex flow regions, which require a higher-order turbulence model, only occur in a small part of the domain. Therefore, the combination of the computational efficiency of the RANS approach with an LES or DNS formulation, promising a higher accuracy, is capable to yield physically more correct results at minimized additional costs compared to pure RANS solutions. An overview of such hybrid RANS-LES approaches is given in [8]. There are at least two widely used techniques to couple RANS with LES. The first approach uses a continuous turbulence model, which switches from RANS to LES to close the system of equations in a unified domain, such as the detached-eddy simulation (DES) proposed by Spalart et al. [24]. The transition from RANS to LES is triggered by the local grid size and the wall distance, which means that where the mesh is fine enough to resolve relevant energy containing eddies, the eddy viscosity of the RANS model is reduced to a subgrid scale model. This approach suffers, however, from a so called grey zone, which occurs when the DES model is already switched into LES mode, but the larger scales of the turbulence spectrum are not established in the solution yet. Therefore, it is difficult to switch the DES model from RANS to LES mode e.g. in an attached boundary layer.

The second technique uses two or more predefined separate computational zones that are linked via an overlapping region, where the transition from RANS to LES and vice versa occurs. In the RANS zone a coarse mesh is sufficient for the solution, while in the LES regions a fine mesh is used to allow the required resolution of the turbulent scales up to the inertial range. The interface conditions between the RANS and LES regimes constitute the major challenge of this second technique which will be denoted zonal technique in the following. For the transition from RANS to LES the information of the turbulent flow of the RANS domain must be used to generate a physically correct turbulence spectrum within the overlapping zone of the RANS and LES domains. That is, the mean velocity distribution of the RANS solution and turbulent fluctuations are imposed at the inflow boundary of the embedded LES domain.

There exist several possibilities to generate such turbulent fluctuations at the inflow boundary [19]. Batten et al. [3] reformulated on the ideas of Kraichnan [14] and Smirnov et al. [21] for wall bounded flows. The velocity signal is generated by a sum of sines and cosines with random phases and amplitudes. The wave numbers are calculated from a three-dimensional spectrum and are scaled by the values of the Reynolds-stress tensor. A special wall treatment was applied to elongate near-wall structures. A transition length to physical turbulence of about ten channel half heights was obtained at low Reynolds number channel flow.

Pamiès et al. [19] expanded the method of Jarrin et al. [10] by dividing the inflow plane of an incompressible flat plate boundary layer into several zones depending on the wall distance. At each zone turbulent eddy shapes are prescribed in the sense of Marusic [17], i.e., these shapes are representative for typical coherent structures of the turbulent boundary layer. This resulted in a good approximation for the low-order statistics of wall-bounded flows and reduced the transition length to approximately five boundary-layer thicknesses without using control planes downstream of the LES inflow boundary. Note that the analysis is focused on an incompressible boundary layer at a very limited Reynolds number range at zero-pressure gradient. Furthermore, the averaged inflow conditions such as averaged velocity profile and Reynolds stress tensor were extracted from a fully developed LES solution that was computed *a priori*.

In this study, the ansatz of Pamiès *et al.* [19] is modified and generalized such that incompressible and compressible flows at a wide Reynolds number range can be computed by a robust and efficient zonal RANS-LES method. The averaged inflow conditions are provided by a RANS simulation and the RANS-to-LES transition behavior is analyzed in detail.

The paper is organized as follows. In Section 2, the numerical flow solver and the synthetic turbulence generation method are described. Subsequently, in Section 3 the flow problems, i.e., the flat-plate flows are introduced. Section 4 contains the results. That is, solutions of the zonal method are compared with DNS and experimental findings. Finally, results for the zonal RANS-LES method are presented for a transonic airfoil flow and some concise conclusions are drawn.

2   Numerical Method

2.1   Flow Solver

The three-dimensional unsteady compressible Navier-Stokes equations are solved based on a large-eddy simulation (LES) using the MILES (monotone integrated LES) approach [4]. The vertex-centered finite-volume flow solver is block-structured. A modified advection-upstream-splitting method (AUSM) is used for the Euler terms [16] which are discretized to second-order accuracy by an upwind-biased approximation. For the non-Euler terms a centered approximation of second-order is used. The temporal integration from time level n to $n+1$ is done by a second-order accurate explicit 5-stage Runge-Kutta method, the coefficients of which are optimized for maximum stability. For a detailed description of the flow solver the reader is referred to Meinke *et al.* [18].

The RANS simulations use the one-equation turbulence model of Fares and Schröder [7] to close the averaged equations.

2.2   Synthetic Turbulence Generation Method

The method used in this paper is based on the work of Jarrin *et al.* [10] and Pamiès *et al.* [19], called synthetic eddy method (SEM), which describes turbulence as a superposition of coherent structures. These structures are generated over the LES inlet plane by superimposing the influence of virtual eddy cores that are defined in a specified volume around the inlet plane that has the streamwise, wall-normal, and spanwise dimensions of the turbulent length-scale l_1, the boundary-layer thickness at inlet δ_0, and the width of the computational domain L_z, respectively. N virtual eddy cores are defined at positions x_m^i inside

of the virtual box and their local influence on the velocity field is defined by
a shape function σ which describes the spatial and temporal characteristics of
the turbulent structure. The normalized stochastical velocity fluctuation components u'_m at the coordinate x_m at the LES inflow plane reads

$$u'_m (x_{1,2,3}, t) = \frac{1}{\sqrt{N}} \sum_{i=1}^{N} \epsilon^i f_{\sigma m} (\tilde{x}_n) \, , \tilde{x}_n = \frac{x_n - x_n^i}{l_n} , \tag{1}$$

where the superscript i denotes a virtual eddy core, ϵ^i the random sign, and
$m, n = 1, 2, 3$ the Cartesian coordinates in streamwise, wall-normal, and spanwise
direction, respectively. . The shape function $f_{\sigma m}$ that has a compact support on
$[-l_n, l_n]$ where l_n is a length scale which satisfies the normalization condition
$\frac{1}{\sqrt{2\pi}} \int_{-1}^{1} f_{\sigma m}^2 \, d\tilde{x}_m = 1$. Jarrin $et\ al.$ used as shape function $f_{\sigma m=1,2,3}$ a Gauss-
or a tent function. The virtual eddy cores convect with the velocity U_{con} in
streamwise direction. Once $x_1^i > l_1$ a new eddy core assigned with randomly
chosen coordinates x_m^i and signs ϵ^i is generated.

The velocity signal at the LES inflow plane is composed of an averaged
velocity component which is in this work provided from the upstream RANS
solution and the normalized stochastic fluctuation u'_m of Eq. 1 that is subjected
to a Cholesky decomposition A_{mn} to assign the values of the Reynolds-stress
tensor R_{mn}.

$$u_m (x, t) = U_{RANS}^m + \sum_n A_{mn} u'_m (x, t) . \tag{2}$$

Pamiès $et\ al.$ [19] extended the method by dividing the inflow plane in several
domains p depending on the distance from the wall. Each domain is characterized
by specific shape factors, turbulent length- and time scales. Thus, the velocity
fluctuation component of Eq. 1 yields

$$u'_m (x_{1,2,3}, t) = \sum_{p=1}^{P} u'_{m,p} (x_{1,2,3}, t) \tag{3}$$

where P denotes the number of divided domains of the inflow plane. Pamiès et
$al.$ defined the shape function $f_{\sigma_p^n}$ of the first two planes according to the educed
turbulent structures of Jeong $et\ al.$ [12],

$$f_{\sigma_{p=1,2}^{m=1}} = G (\tilde{x}_1) G (\tilde{x}_2) H (\tilde{x}_3)$$
$$f_{\sigma_{p=1,2}^{m=2}} = -G (\tilde{x}_1) G (\tilde{x}_2) H (\tilde{x}_3)$$
$$f_{\sigma_{p=1,2}^{m=3}} = G (\tilde{x}_1) H (\tilde{x}_2) G (\tilde{x}_3)$$

where $H (\tilde{x}_m) = 1 - \cos (2\pi \tilde{x}_m) / (2\pi \cdot 0.44)$ and $G (\tilde{x}_m)$ is a Gaussian function.

In this work the inflow plane was divided in three planes, that is $P = 3$. The
position in wall-normal direction $x_{2,beg}, x_{2,end}$ of each plane p and the corresponding length scales in streamwise, wall-normal, and spanwise direction, and
convection velocities are given in Tab. 1. The length scales of the turbulent structures l_n in the first plane $p = 1$ are chosen accordingly to Pamiès $et\ al.$ [19] and

del Alamo *et al.* [2]. However, the length scales of the structures in the second and third plane $p = 2, 3$ are set to values that are different compared to Pamiès *et al.* The analysis of several incompressible and compressible boundary layers at various Reynolds numbers has shown that the values chosen by Pamiès *et al.* at $p = 2, 3$ did not satisfactorily match the reference flow field.

The shear-stress component $\langle u_1' u_2' \rangle$ of the Reynolds-stress tensor R_{mn} that is needed for Eq.2 is obtained from the RANS solution located upstream of the LES inlet [20]. The normal-stress components are reconstructed using a fourth order polynomial function to match the distribution of Spalart [1].

Morkovin' s hypothesis is applied at the inlet to relate density and velocity fluctuations and to enforce the strong Reynolds analogy (SRA) [22]. The density field is obtained by enforcing a constant-pressure condition at the inflow [6].

Table 1. Locations of planes p, turbulent length scales l_n, and convection velocities U_{con}

plane	$l_{y,p}$ $[x_{2,beg}; x_{2,end}]$	$= l_1$	l_2	l_3	U_{con}
$p = 1$	$[0; (60)^+]$	$(100)^+$	$(20)^+$	$(60)^+$	$0.6U_\infty$
$p = 2$	$[(60)^+; 0.65\delta_0]$	$0.5\delta_0$	$0.3\delta_0$	$0.25\delta_0$	$0.75U_\infty$
$p = 3$	$[0.65\delta_0; 1.2\delta_0]$	$0.3\delta_0$	$0.3\delta_0$	$0.3\delta_0$	$0.9U_\infty$

3   Computational Setup

Flat Plate Boundary Layer. A subsonic flat-plate boundary-layer flow is investigated to validate the STG method for the zonal RANS-LES configuration comparing the results with a pure RANS, pure LES, and available experimental data. The freestream Mach numbers are $M = 0.4$ and $M = 2.3$ and the freestream Reynolds numbers based on the momentum thickness at $x/\delta_0 = 0$ are $Re_\theta = 1400$ and $Re_\theta = 4200$, respectively. where δ_0 denotes the boundary-layer thickness at the inlet of computational domain of the pure LES, pure RANS, and the embedded LES part of the zonal RANS-LES simulation. The inflow boundaries of the pure LES, pure RANS, and the embedded LES part of the zonal RANS-LES simulation are located at $x/\delta_0 = 0$.

The numerical details of each simulation are presented in Tab. 2. The grids are clustered to the surface in the wall-normal direction using a hyperbolic tangent stretching function such that the minimum grid spacing in wall units is approximately one and a stretching factor of 1.05 is not exceeded. Depending on the configuration subsonic and supersonic outflow boundary conditions are used at the upper and downstream boundaries. The no-slip boundary condition is imposed at the adiabatic wall. The inflow distribution of the flow variables for the LES inlet of the zonal RANS-LES simulation were extracted from the RANS part that is located upstream of the LES domain.

Table 2. Computational domain, grid resolution, and number of mesh points for pure LES, pure RANS, and zonal RANS-LES configurations of turbulent boundary layer simulations. The zonal RANS-LES configuration consists of the RANS domains upstream (Zo-RANS) and of the embedded LES domain (Zo-LES).

domain	domain size			resolution			number of grid points		
	L_x/δ_0	L_y/δ_0	L_z/δ_0	Δx^+	Δy^+_{wall}	Δz^+	i_{max}	j_{max}	k_{max}
pure LES ($M = 0.4$)	16.0	3.4	0.88	15.1	1.1	6.7	516	67	49
pure 2D-RANS ($M = 0.4$)	16.0	3.4	-	62.2	1.1	-	104	67	-
Zo-RANS ($M = 0.4$)	4.0	3.4	0.88	61.1	1.1	160	31	67	3
Zo-LES ($M = 0.4$)	12.0	5.0	0.88	15.1	1.1	6.7	387	67	49

The inflow distributions of the pure LES results is determined using the rescaling method of El-Askary *et al.* [6]. The recycling station is located at $x/\delta_0 = 6$. A sponge layer is applied at the upper- and outflow boundary to damp spurious pressure fluctuations. The wall- and velocity outflow boundary conditions are the same as for the formulations of the pure 2D-RANS configuration.

Transonic Airfoil Flow. The transonic flow around a DRA2303 airfoil [9] was chosen as the aerodynamic reference case to discuss the efficiency and quality of the zonal RANS-LES method compared to a pure LES method. The flow field is defined by $M = 0.72$, $Re_c = 2.6 \cdot 10^6$ based on the chord length c, and the angle of attack $\alpha = 3°$. The laminar-turbulent transition is fixed at the pressure and suction side of the airfoil at $x/c = 0.05$ for both numerical configurations by introducing a wall surface roughness of an amplitude of approximately 10 inner wall units or $8 \cdot 10^{-4} \Delta y/c$.

Table 3. Computational domain, grid resolution, and number of mesh points for pure LES and zonal RANS-LES configuration for the transonic airfoil case. The zonal RANS-LES configuration consists of the RANS domains (Zo-RANS) and of the embedded LES domain (Zo-LES).

domain	domain size		resolution			number of grid points			
	$L_{farfield}$	$L_{spanwise}$	Δx^+	Δy^+_{wall}	Δz^+	I_{max}	j_{max}	k_{max}	total
pure LES	25c	0.021c	100	1.0	20	2364	130	97	$30 \cdot 10^6$
Zo-RANS	25c	0.021c	400	1.0	180	225	89	11	$2.2 \cdot 10^5$
Zo-LES	0.4c	0.021c	100	1.0	20	1430	97	97	$13.5 \cdot 10^6$

The resolution of the pure LES grid in the streamwise, wall normal and spanwise direction of $\Delta x^+ \approx 100$, $\Delta y^+_{min} \approx 1$, and $\Delta z^+ \approx 20$, respectively, yields a total number of grid points of approximately $30 \cdot 10^6$. The spanwise extension of the grid is 0.021 c. Using the same grid resolution and spanwise extension,

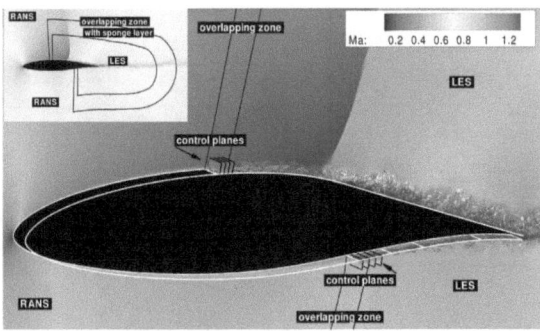

Fig. 1. Computational configuration of the zonal RANS-LES computation and Mach number contours. In the LES zone λ_2-contours [11] are shown color coded with mapped-on local Mach number.

the number of grid points of the embedded LES domain of the zonal RANS-LES configuration is approximately $13.7 \cdot 10^6$, i.e., the reduction is more than a factor of two. Details of the grid configurations are given in Tab. 3. The pure LES uses periodic boundary conditions in the spanwise direction and a no-slip, adiabatic condition is imposed on the wall. Non-reflective boundary conditions are applied to the far field boundaries. The computational setup of the zonal RANS-LES computation is shown in Fig. 1. The zonal RANS-LES configuration uses the same boundary conditions at the wall and in the far field as the pure LES computation. At the inflow boundary of the LES domain on the upper and lower side of the airfoil, the STGM discussed in Sec. 2 is applied to generate synthetic turbulent structures in the turbulent boundary layer. Downstream of the LES inflow boundary at the upper side four control planes are located between $0.37 \leq x/c \leq 0.4$ and at the lower side between $0.7 \leq x/c \leq 0.73$. The time-averaged velocity profile and the Reynolds shear stress component $\langle u'v' \rangle$ of the upstream RANS solution are used as target conditions for the STGM and the control planes. At the RANS outflow the time-averaged pressure from the embedded LES domain located downstream is prescribed whereas density and velocity distributions are extrapolated. At the LES inflow the density and velocity distributions from the upstream RANS domain are imposed and the pressure values are extrapolated from the interior of the embedded LES domain. The LES domain is encompassed by a sponge layer to damp spurious pressure fluctuations.

4 Results

Subsonic Boundary Layer. In this section, the findings of the subsonic flat-plate boundary layer flow applying the STG method for the zonal RANS-LES ansatz are discussed. In the subsequent paragraphs the term zonal RANS-LES is applied for the results of the corresponding embedded LES domain. In Sec. 4 the inflow method is validated for a subsonic flat-plate boundary-layer flow,

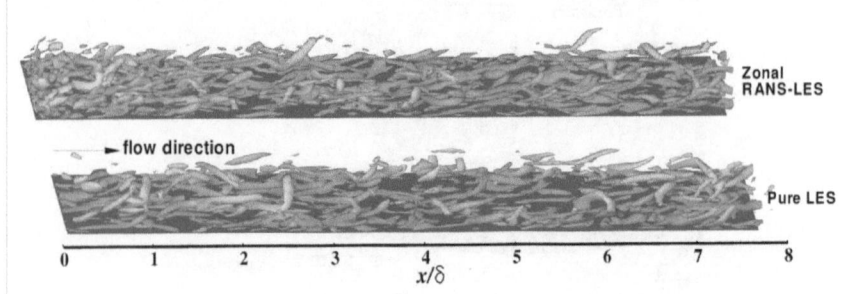

Fig. 2. Coherent turbulent structures based on the λ_2-criterion with mapped-on local Mach number for subsonic flat-plate boundary layer.

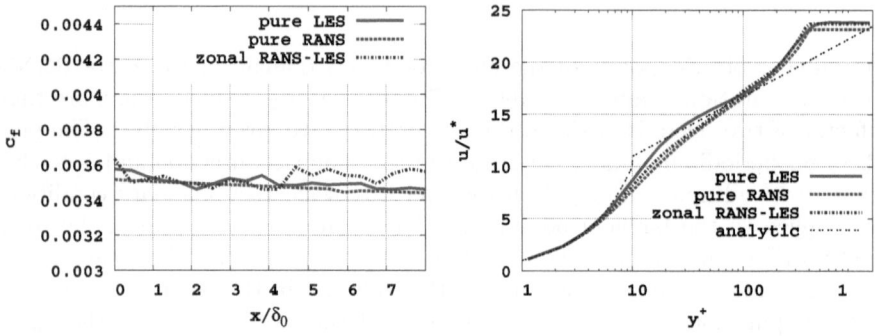

Fig. 3. Skin-friction distributions (left) and van-Driest-velocity distributions at $x/\delta_0 = 2$(right) for several numerical configurations.

respectively, by comparing the averaged boundary-layer properties and turbulent flow field with reference LES, and experimental data. The development of the coherent turbulent structures in the pure LES and zonal RANS-LES solution is discussed and the streamwise distributions of the skin-friction coefficient c_f, the shape factor H, and the displacement thickness δ_1 of the pure LES, and the zonal RANS-LES solution are compared. For the subsonic case the Reynolds shear stress distributions at $x/\delta_0 \approx 2$ of the zonal RANS-LES are compared with pure LES and measurements of deGraaff and Eaton [5].

Coherent turbulent structures based on the λ_2-criterion according to Jeong and Hussain [11] with mapped-on Mach number contours are visualized in Fig. 4 for the zonal RANS-LES solution and the pure LES. Near the inflow boundary of the LES domain of the zonal RANS-LES simulation at $x/\delta_0 < 1$ elongated structures are already visible. At $x/\delta_0 > 1$ the size and number of those structures is comparable to that of the pure LES result. The STG method presented in Sec. 2.2 generates coherent turbulent structures that contain the appropriate length- and time scales which form flow patterns downstream of the inlet that

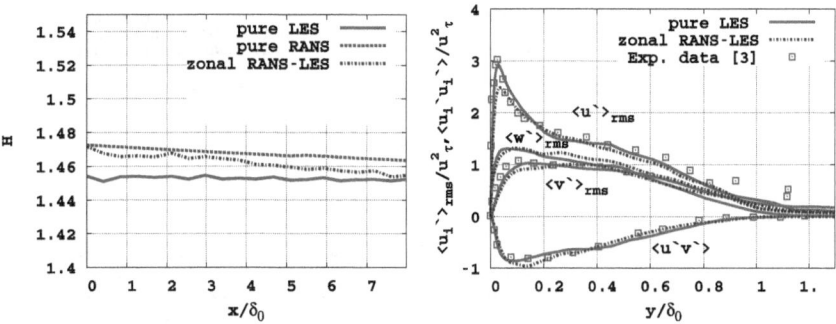

Fig. 4. Streamwise development of the shape-factor (left) for several numerical configurations and comparison of Reynolds normal-stress component distributions of pure LES, zonal RANS-LES, and reference experimental results [5] at $x/\delta_0 = 2$ (right).

resemble the turbulent structures of the pure LES solution. That is, already at $x/\delta_0 \approx 1$ ejected vortices are observed and elongated structures in the streamwise direction that are essential for the turbulence production develop further downstream.

The streamwise development of the skin-friction coefficient c_f is presented in Fig. 3(a). The c_f-distribution for the pure RANS and the zonal RANS-LES results are in good agreement with the pure LES solution. Downstream of the LES inflow of the zonal RANS-LES the skin-friction coefficient does not drop but rather immediately converges to the pure LES values. The structures generated by the original inflow method of Jarrin *et al.* [10] would too strongly dissipate such that a much larger streamwise extent would be necessary for the LES to recover the correct c_f-level.

In Fig. 3(b) the van-Driest velocity distribution at x/δ_0 of pure LES, pure RANS and the zonal RANS-LES simulation is shown. The distribution of the zonal RANS-LES resembles that of the pure RANS, however, it started to converge to the distribution of the pure LES.

Figure 4(a) shows the time-averaged streamwise distribution of the shape factor H. The growth rates of the pure RANS, the pure LES, and the zonal RANS-LES simulation are more or less alike. From the streamwise distributions of the skin-friction coefficient and the displacement thickness it can be concluded that the zonal RANS-LES method yields smooth streamwise results which are comparable with the pure LES findings.

The distributions of the Reynolds normal- and shear-stress components of the pure LES and the zonal RANS-LES configuration are compared with the experimental results $Re_\theta = 1430$ of deGraaff and Eaton [5] in Fig. 4(b). A good agreement with the experimental data is obtained corroborating that the inflow generation method for the zonal RANS-LES configuration is capable of generating physically meaningful Reynolds stresses within a short transition length, i.e., in less than two boundary-layer thicknesses δ_0.

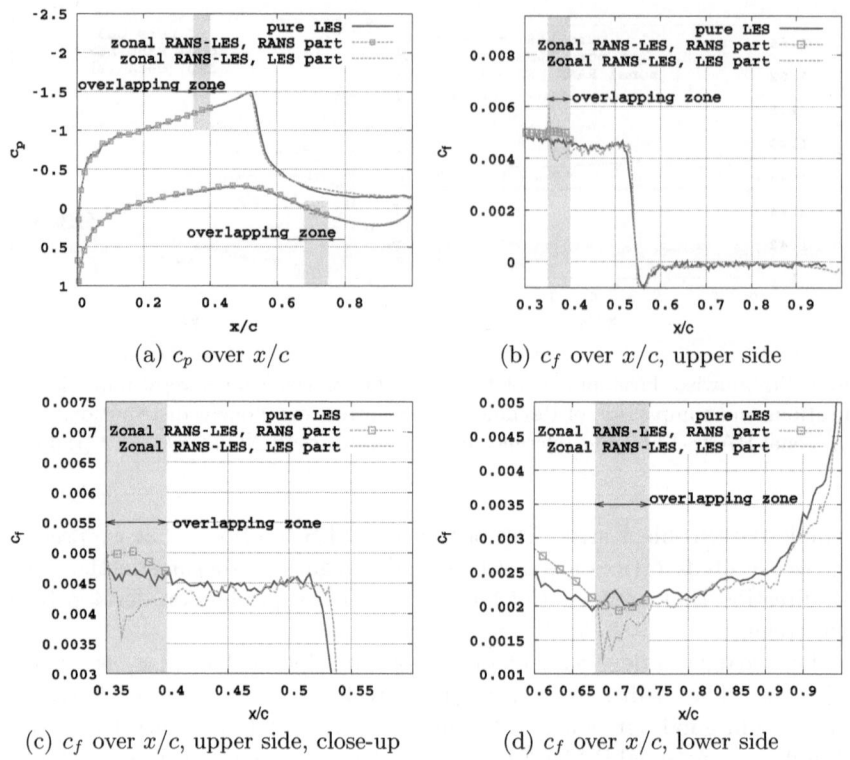

(a) c_p over x/c

(b) c_f over x/c, upper side

(c) c_f over x/c, upper side, close-up

(d) c_f over x/c, lower side

Fig. 5. Pressure coefficient distribution c_p and skin-friction coefficient distribution c_f at the upper and lower side of the DRA2303 airfoil for the zonal RANS-LES and the pure LES.

Transonic Airfoil Flow. In Fig. 5(a) the time- and spanwise averaged distributions of the pressure coefficient c_p for the zonal RANS-LES and the pure LES are presented. The averaging time was about two shock-oscillation cycles. The gray shaded areas represent the overlapping regions of the zonal RANS-LES approach. The average shock position is located at $x/c \approx 0.57$ for the zonal RANS-LES and the pure LES result. A smooth RANS-to-LES transition of the pressure coefficient at the upper and lower side of the airfoil is evident.

The skin-friction coefficient distributions at the upper side of the airfoil are presented in Fig. 5(a). The c_f distribution of the zonal RANS-LES agrees well over the entire upper side of the airfoil with the pure LES result. From the shock position at $x/c \approx 0.57$ to the trailing edge the averaged flow field is fully separated.

In Fig. 6(a) the velocity distribution of the zonal RANS-LES and the pure LES solutions are compared at $x/c = 0.50$ which is located upstream of the average shock position at $x/c \approx 0.57$. A slight deviation near the boundary-layer edge in

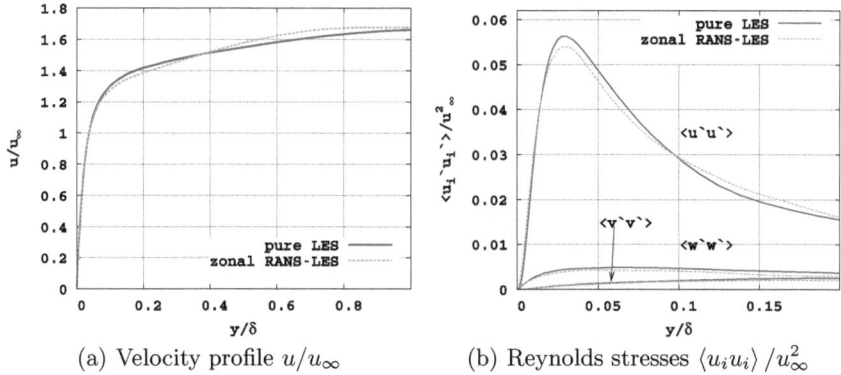

(a) Velocity profile u/u_∞ (b) Reynolds stresses $\langle u_i u_i \rangle /u_\infty^2$

Fig. 6. Velocity profile and normal components of the Reynolds-stress tensor at $x/c = 0.50$ at the upper side of the DRA2303 airfoil for the zonal RANS-LES and the pure LES.

the velocity distribution is observed. However, near the wall the difference between the velocity profiles is small resulting in almost identical c_f-values.

The distributions of the normal components of the Reynolds stress tensor at $x/c = 0.50$ for the zonal RANS-LES and the pure LES computations are shown in Fig. 6(b). The normal stresses computed by the zonal RANS-LES method are in very good agreement with the pure LES results. This convincing match of the velocity and the Reynolds-normal-stress distributions constitute a crucial requirement to obtain similar shock dynamics as well as time-averaged shock positions.

5   Conclusion

A synthetic turbulence generation method for a zonal RANS-LES method for sub- and supersonic flows has been introduced. The STG method has been validated by computing a subsonic boundary-layer flow at $M = 0.4$ and $Re_\theta = 1400$ and a supersonic flow boundary-layer flow at $M = 2.3$ and $Re_\theta = 4200$, respectively. The zonal RANS-LES solutions were compared with pure LES, pure RANS, DNS, and experimental data. A rapid RANS-to-LES transition was observed and the overall accuracy has been convincing. Within a transition length from the RANS to the LES solution of approximately two boundary-layer thicknesses the zonal ansatz showed good agreement in the streamwise c_f distribution, the velocity profiles, and the distribution of the Reynolds stresses compared with measurements [5]. Also the growth rate of boundary-layer-shape factor, the boundary-layer-displacement thickness in the streamwise direction of the zonal RANS-LES solution was in good agreement with that of the pure LES results.

The convincing agreement of the zonal RANS-LES results with the pure LES solutions for the transonic airfoil flow increases the confidence in the application of the zonal RANS-LES method. Since no modifications of the interface formulations are necessary it is more or less straightforward to apply the zonal RANS-LES method to other three-dimensional sub- and transonic flow problems.

References

1. Spalart, P.R.: Direct simulation of a turbulent boundary layer up to $Re_\theta = 1410$. J. Fluid Mech. **187**, 61–98 (1988)
2. del Alamo, J.C., Jimenez, J., Zandonade, P., Moser, R.D.: Self-similar vortex clusters in the turbulent logarithmic region. J. Fluid Mech. **561**, 329 (2006)
3. Batten, P., Goldberg, U., Chakravarthy, S.: Interfacing statistical turbulence closures with large-eddy simulation. AIAA J. **42**(3), 485–492 (2004)
4. Boris, J.P., Grinstein, F.F., Oran, E.S., Kolbe, R.L.: New insights into large eddy simulation. Fluid Dynamics Research **10**, 199–228 (1992)
5. DeGraaff, D.B., Eaton, J.K.: Reynolds-number scaling of the flat-plate turbulent boundary layer. J. Fluid Mech. **422**, 319–346 (2000)
6. El-Askary, W., Schröoder, W., Meinke, M.: LES of compressible wall-bounded flows. In: AIAA Paper (2003–3554) (2003)
7. Fares, E., Schröoder, W.: A general one-equation turbulence model for free shear and wallbounded flows. Flow, Turbulence and Combustion **73**, 187–215 (2004)
8. von Fröhlich, J., Terzi, D.: Hybrid LES/RANS methods for the simulation of turbulent flows. Prog. Aerospace Sci. **44**, 349–377 (2008)
9. Fulker, J.L., Simmons, M.J.: An Experimental Investigation of Passive Shock/Boundary Layer Interaction Control on an Aerofoil. Draiasihwaicr 9521611 EUROSHOCK Tr Aer 2 4913(2) (1992)
10. Jarrin, N., Benhamadouche, S., Laurence, D., Prosser, R.: A synthetic-eddy-method for generating inflow conditions for large-eddy simulations. Int. J. Heat Fluid Flow **27**, 585–593 (2006)
11. Jeong, J., Hussain, F.: On the identification of a vortex. J. Fluid Mech. **285**, 69–94 (1995)
12. Jeong, J., Hussain, F., Schoppa, W., Kim, J.: Coherent structures near the wall in a turbulent channel flow. J. Fluid Mech. **332**, 185 (1997)
13. Knight, D.D., Yan, H., Panaras, A.G., Zheltovodov, A.A.: Advances in CFD prediction of shock wave turbulent boundary layer interactions. Progress in Aerospace Science **39**, 121–184 (2003)
14. Kraichnan, R.H.: Inertial Ranges in Two-Dimensional Turbulence. Phys. Fluids **10**(7), 1417–1423 (1967)
15. Leschziner, M., Drikakis, D.: Turbulence modelling and turbulent-flow computation in aeronautics. Aeronautical Journal **106**(1061), 349–383 (2002)
16. Liou, M.S., Steffen, C.J.: A new flux splitting scheme. Journal of Computational Physics **107**, 23–39 (1993)
17. Marusic, I.: On the role of large-scale structures in wall turbulence. Physics of Fluids **13**, 735 (2001)
18. Meinke, M., Schröder, W., Krause, E., Rister, T.: A comparison of second- and sixth-order methods for large-eddy simulations. Computers and Fluids **31**, 695–718 (2002)
19. Pamiès, M., Weiss, P.E., Garnier, E., Deck, S., Sagaut, P.: Generation of synthetic turbulent inflow data for large eddy simulation of spatially evolving wall-bounded flows. Physics of Fluids **16**, 045103 (2009)

20. Roidl, B., Meinke, M., Schröder, W.: Zonal RANS/LES computation of transonic airfoil flow. In: AIAA Paper (2011–1056) (2011)
21. Smirnov, A., Shi, S., Celik, I.: Random flow generation technique for large eddy simulations and particle dynamics modeling. J. Fluids Eng. **123**, 359–371 (2001)
22. Smits, A.J., Dussauge, J.P.: Turbulent Shear Layers in Supersonic Flow, 2nd edn. Springer, New York (2006)
23. Spalart, P.: Direct simulation of a turbulent boundary layer up to req = 1410. J. Fluid Mech. **187**, 61–98 (1988)
24. Spalart, P.R., Jou, W.H., Strelets, M., Allmaras, S.R.: Comments on the feasibility of LES for wings, and on a hybrid RANS/LES approach. In: Advances on DNS/LES, pp. 137–147. Greyden Press, Columbus, OH (1997)

Mathematical Algorithm for Calculation of the Moving Tsunami Wave Height

Sergey Kabanikhin[1] and Olga Krivorotko[2(✉)]

[1] Institute of Computational Mathematics and Mathematical Geophysics SB RAS,
Prospect Akademika Lavrentjeva, 6, 630090 Novosibirsk,
Russian Federation
kabanikhin@sscc.ru
http://www.sscc.ru/index_e.html
[2] Novosibirsk State University, Pirogova Str., 2, 630090 Novosibirsk,
Russian Federation
krivorotko.olya@mail.ru
http://www.nsu.ru

Abstract. New numerical algorithm of determining the moving tsunami wave height for linear source at the characteristic surface $t = \tau(x, y)$ is proposed where $\tau(x, y)$ is a solution of Cauchy problem for eikonal equation. This algorithm based on and representation of fundamental solution of linear shallow water equations in the singular and regular parts. This approach allows one to reduce computational time. We get the expression of the moving tsunami wave height for the linear and arbitrary sources. Numerical results are discussed.

Keywords: Shallow water equations · Wave front amplitude · Impulse source · Eikonal equation · Finite difference method

1 Introduction

The recent severe tsunamis in Japan (2011), Sumatra (2004), and at the Indian coast (2004) showed that a system producing exact and immediate information about tsunamis is of vital importance. Mathematical modeling and numerical simulations are most used instruments for providing a such information. Most suitable physical models related to simulation of tsunamis are based on linear shallow water equations:

$$\begin{cases} \eta_{tt} = \operatorname{div}(gH(x,y)\operatorname{grad}\eta), & t \in (0, T); \\ \eta|_{t=0} = q(x, y), & \eta_t|_{t=0} = 0, \ (x, y) \in \Omega. \end{cases} \quad (1)$$

Here $\Omega := (0, L_x) \times (0, L_y)$ is a rectangle domain, $\eta(x, y, t)$ is the free surface, $H(x, y) > 0$ is a known function describing the bottom relief (bathymetry),

O. Krivorotko–This work is partially supported by the Ministry of Education and Science of the Russian Federation and the Republic of Kazakhstan N. 1746/GF4 "Theory and numerical methods for solving inverse and ill-posed problems of nature".

N. Danaev et al. (Eds.): CITech 2015, CCIS 549, pp. 66–72, 2015.
DOI: 10.1007/978-3-319-25058-8_7

$q(x, y)$ is an initial tsunami perturbation in Ω, $g = 9.8\ [m/s^2]$ is the acceleration of gravity. Assume that the time period T is not long enough for the wave to reach the edges of the domain, and therefore we can set homogeneous boundary conditions at the boundary $\partial\Omega$ of the domain, $i.e.$ $\eta|_{\partial\Omega} = 0$.

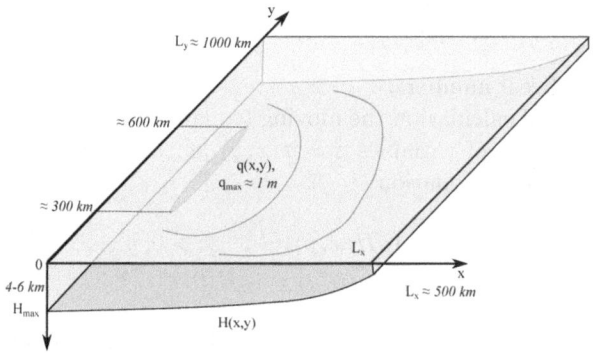

Fig. 1. Illustration of the calculation domain Ω.

Simulation of tsunami wave propagation on such scales $(10^6\ \mathrm{kms}^2$ on space and about one hour when initial wave amplitude is no more than 2 meters) is not an easy calculation task (see Figure 1) [1].

V.M. Babich ([2], § 5) developed space-time ray approach for getting expression of tsunami wave front amplitude in case of "slowly varying" bathymetry with caustics. In paper [3] asymptotic method for determining of tsunami wave front amplitude based on the generalization of the construction known as the Maslov canonical operator [4] was proposed. Practical application of some algorithms can be found in papers by L.B. Chubarov, Yu.I. Shokin et al. [5], [6], [7] and website http://www.ict.nsc.ru. Our numerical algorithm makes it possible to calculate the front amplitude of a wave coming to a given point and the wave arrival time by solving problem not in the entire domain Ω, but only on a selected characteristic surface $t = \tau(x, y)$. Here $\tau(x, y)$ is a solution of a Cauchy problem for eikonal equation $\tau_x^2 + \tau_y^2 - (gH(x, y))^{-1}$, $\tau(0, y) = 0$.

The paper is organized as follows. In Section 2 the Cauchy problem for determining of moving tsunami wave height is derived in case of linear tsunami source. Numerical experiments for linear source are described in Section 2.1. Obtained results correlate with well-known Airy-Green formula (see Section 2.2). In Section 3 using characteristics of hyperbolic equation we demonstrate the formula for tsunami wave height for point source. In the last Section 4 we show connections between moving tsunami wave height for point, linear and arbitrary sources and denote the plans for future work.

2   Wave Front Amplitude for Linear Source

We consider the linear source $q(x, y) = h(y) \cdot \delta(x)$. Here $\delta(x)$ is a Dirac function and $h(y)$ is a sufficiently smooth function. Then problem (1) can be reduced to the following problem in a half-plane

$$\begin{cases} \eta_{tt} = \operatorname{div}(gH(x,y)\operatorname{grad}\eta), & x > 0, y \in \mathbb{R}, t \in \mathbb{R}, \\ \eta|_{t<0} = 0, \quad \eta_x|_{x=0} = h(y) \cdot \delta(t), \ y \in \mathbb{R}, t \in \mathbb{R}. \end{cases} \quad (2)$$

Here $\mathbb{R}$ is a set of real numbers.

The main idea of calculation the moving tsunami wave height for problem (2) consists of the change of variables $z = \tau(x, y)$ [8]. Here $\tau(x, y)$ is a solution of the problem for eikonal equation

$$\begin{cases} \tau_x^2 + \tau_y^2 = (gH(x,y))^{-1}, & x > 0, y \in \mathbb{R}; \\ \tau(0,y) = 0, \ \tau_x(0,y) = (gH(0,y))^{-1/2}. \end{cases} \quad (3)$$

Remark. We assume that maps $z = \tau(x, y)$ and $x = x(z, y)$ are mutually inverse and one-to-one, *i.e.* we exclude the appearance of caustics.

Then, with the change of variables $v(z, y, t) = \eta(x, y, t)$ and $b(z, y) = \sqrt{gH(x,y)}$ problem (2) can be rewritten as follows ($z, y > 0$):

$$\begin{cases} v_{tt} = v_{zz} + b^2 v_{yy} + A_1 v_{zy} + A_2 v_z + A_3 v_y, \ z > 0, y \in \mathbb{R}, t \in \mathbb{R}, \\ v|_{t<0} = 0, \quad v_z|_{z=0} = g(y)\delta(t), & y \in \mathbb{R}, t \in \mathbb{R}. \end{cases} \quad (4)$$

$A_1 = 2b^2 \tau_y$, $A_2 = b^2(\tau_{xx} + \tau_{yy}) + 2(\frac{b_z}{b} + bb_y \tau_y)$, $A_3 = 2b(b_z \tau_y + b_y)$, $g(y) = h(y)\left(b^{-2}(0,y) - \tau_y^2(0,y)\right)^{-1/2}$.

The coefficients of derivatives v_{tt} and v_{zz} in problem (4) are equal to unity that allows us to represent the solution of problem (4) as follows [9]:

$$v(z, y, t) = S^{(l)}(z, y) \cdot \theta(t - z) + \tilde{v}(z, y, t). \quad (5)$$

Here $\tilde{v}(z, y, t)$ is a smooth function, $\theta(t - z)$ is a Heaviside step function.

Substituting representation (5) in system (4) and equating the coefficients at $\delta(t - z)$, we obtain a Cauchy problem for wave amplitude $S^{(l)}(z, y)$:

$$\begin{cases} S_z^{(l)} + 0.5 A_1 S_y^{(l)} + 0.5 A_2 S^{(l)} = 0, \ z > 0, y \in \mathbb{R}; \\ S^{(l)}(0, y) = g(y), & y \in \mathbb{R}. \end{cases} \quad (6)$$

The main benefit of above algorithm consists of reducing of the problem (1) to the problem (6) of determining the function $S^{(l)}(z, y)$ of two variables.

2.1   Numerical Experiment

In numerical calculations we use 1:10000 m scale. We put $L_x = 70$ km and $L_y = 100$ km. The initial wave amplitude is equal to 1 m. The grid size is equal to 500×300 points. We solve the problem (3) for artificial bathymetry using

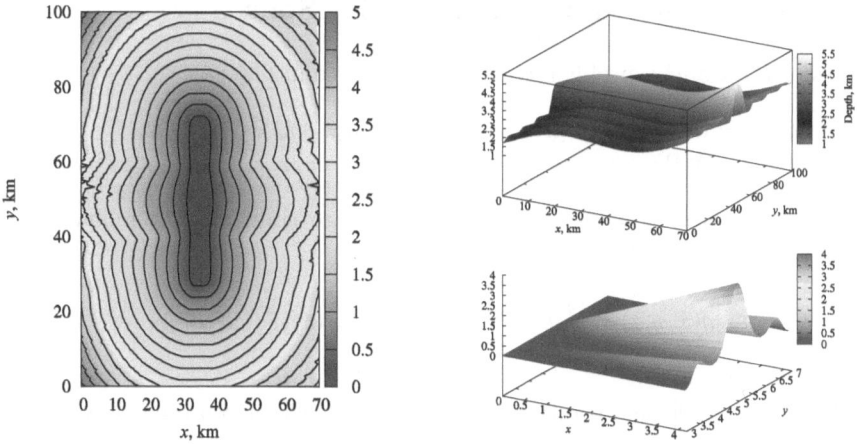

Fig. 2. (*On the left*) The times of arrival of the first waves amplitude from the linear source (space values are given in kilometers, time is given in minutes). (*On the right top*) The bathymetry $H(x, y)$ (all dimensions are in kilometers). (*On the right bottom*) The wave amplitude $S^{(l)}(x, y)$ (amplitude values are given in meters).

a method of characteristics [10] and the Cauchy problem (6) using an explicit finite-difference scheme of the second order approximation (Fig. 2).

Our algorithm allows one to compute an initial wave amplitude using measured data at moment T. Note, that the inversion method of reconstructing of an initial tsunami source $q(x, y)$ from measurements of water-level data (marigrams) was first proposed by T.A. Voronina and V.A. Tcheverda in 1998 [12] and was already described in previous papers [13], [14]. This method based on singular value decomposition approach. The optimization method of reconstructing of an initial tsunami source $q(x, y)$ was proposed in [1]. All these methods deal with solving the shallow water system in the entire domain Ω. Proposed algorithm is 30 times faster than solving complete shallow water equations.

2.2 Airy-Green Formula

If all functions in Cauchy problem (6) does not depend on the variable y (one-dimensional case), then the problem (6) has a following solution [11]:

$$S^{(l)}(z) = S^{(l)}(0) \cdot \sqrt[4]{H(0)/H(z)}. \tag{7}$$

The expression (7) for the moving tsunami wave height is consistent with the well-known Airy-Green formula: the wave amplitude $S^{(l)}$ increases as a depth of the bottom H decreases.

3   Wave Front Amplitude for Point Source

Consider the Cauchy problem with a point source in (x_0, y_0)

$$\begin{cases} w_{tt} = \text{div}(gH(x,y)\text{grad}\,w) + \delta(x - x_0, y - y_0)\delta(t), \ t \in \mathbb{R}; \\ w|_{t<0} = 0, \qquad\qquad\qquad\qquad\qquad\qquad\qquad (x,y) \in \mathbb{R}^2. \end{cases} \quad (8)$$

Remark. Solutions of problems (1) and (8) are connected as follows ($\xi_1^0 = \xi_1 + x_0$, $\xi_2^0 = \xi_2 + y_0$):

$$\eta(x, y, t) = \int\limits_{\mathbb{R}^2} w_t(x - \xi_1, y - \xi_2, t) q(\xi_1^0, \xi_2^0)\, d\xi_1 d\xi_2.$$

It is known [10], that solution w of problem (8) can be represented as follows:

$$w(x, y, t) = \frac{S^{(p)}(x, y; \mathbf{x}_0)}{\sqrt{t^2 - \tau^2(x, y; \mathbf{x}_0)}}\theta(t)\theta(t^2 - \tau^2(x, y; \mathbf{x}_0)) + \tilde{w}(x, y, t).$$

Here $\tau(x, y; \mathbf{x}_0)$ is a solution of eikonal equation (3) with condition $\tau(x, y; \mathbf{x}_0) = O(|\mathbf{x} - \mathbf{x}_0|)$, $\mathbf{x} \to \mathbf{x}_0$ (here $\mathbf{x} = (x, y)$, $\mathbf{x}_0 = (x_0, y_0)$), and the wave amplitude $S^{(p)}(x, y; \mathbf{x}_0)$ has the form:

$$S^{(p)}(x, y; \mathbf{x}_0) = \frac{1}{\pi g H(x_0, y_0)} \frac{\sqrt{\tau(x, y; \mathbf{x}_0)}}{\exp\{0.5gI(\tau(x, y; \mathbf{x}_0))\}}, \quad (9)$$

where $I(\tau(x, y; \mathbf{x}_0)) = \int\limits_0^\tau ((H\tau_x)_x + (H\tau_y)_y)\, d\tau.$

We obtain the expression (9) for the moving tsunami wave height $S^{(p)}(x, y; \mathbf{x}_0)$ for a point source using the Theorem from [10].

4   Connections and Conclusion

Denote by $S(x, y)$ the moving tsunami wave height generated by an arbitrary source $q(x, y)$. Then the amplitudes $S(x, y)$, $S^{(l)}(x, y)$ and $S^{(p)}(x, y)$ for the arbitrary, linear and point sources, respectively, are connected as follows:

$$S(x, y) = \int\limits_{\mathbb{R}^2} q(\xi, \zeta) S^{(p)}(\xi, \zeta)\, d\xi d\zeta,$$

$$S^{(l)}(x, y) = \int\limits_{\mathbb{R}} h(\zeta) S^{(p)}(x, \zeta)\, d\zeta,$$

$$S(x, y) = \int\limits_{\mathbb{R}} p(\xi) S^{(l)}(\xi, y)\, d\xi, \quad q(x, y) = p(x)h(y).$$

We plan to apply above algorithm for real bathymetry using databases collected by non-profit organization WAPMERR (World Agency of Planetary Monitoring and Earthquake Risk Reduction) in modern GIS technology ITRIS (Integrated Tsunami Research and Information System) [15]. WAPMERR has a historical database of alleged tsunami sources around the world which is based on the information about seaquakes, a database of observations of the tsunami waves in coastal areas and bathymetry data.

We propose a new numerical algorithm of determining the moving tsunami wave height for linear source at the characteristic surface $t = \tau(x, y)$, where $\tau(x, y)$ is a solution of eikonal equation. This algorithm based on and representation of fundamental solution of linear shallow water equations in the singular and regular parts. This approach allows one to reduce computational time. We get the expression of the moving tsunami wave height for the linear and arbitrary sources and discuss numerical results.

Proposed algorithm for calculation of the moving tsunami wave height can be used for solving inverse problem of reconstruction of a tsunami source using additional information of tsunami amplitude at $t = T$ [16]. In this case the Cauchy problem (6) can be solved with a reverse time, *i.e.* the usage of the initial condition $S(T, y) = f(y)$ instead $S(0, y) = g(y)$.

References

1. Kabanikhin, S., Hasanov, A., Marinin, I., Krivorotko, O., Khidasheli, D.: A variational approach to reconstruction of an initial tsunami source perturbation. Appl. Numer. Math. **83**, 22–37 (2014)
2. Babich, V.M., Buldyrev, V.S., Molotkov, I.A.: Space-Time Ray Method: Linear and Nonlinear Waves (in Russian). Leningrad University Publisher, Leningrad (1985)
3. Dobrokhotov, S.Y., Nekrasov, R.V., Tirozzi, B.: Asymptotic solutions of the linear shallow-water equations with localized initial data. J. Eng. Math. **69**, 225–242 (2011)
4. Maslov, V.P.: The Complex WKB Method for Nonlinear Equations I: Linear Theory. Birkhäuser Basel, Basel, Boston, Berlin (1994)
5. Shokin, Y.I., Chubarov, L.B., Fedotova, Z.I., Gusyakov, V.K., Babailov, V.V., Beisel, S.A., Eletsky, S.V.: Mathematical modelling in application to regional tsunami warning systems operations. In: Notes on Numerical Fluid Mechanics and Multidisciplinary Design, vol. 101, pp. 52–68 (2008)
6. Beisel, S.A., Chubarov, L.B., Didenkulova, I., Kit, E., Levin, A., Pelinovsky, E., Shokin, Y.I., Sladkevich, M.: The 1956 Greek Tsunami Recorded at Yafo (Israel) and Its Numerical Modeling. J. Geophys. Res. **114**, C09002 (2009)
7. Dutykh, D., Mitsotakis, D., Chubarov, L.B., Shokin, Y.I.: On the contribution of the horisontal sea-bed displacements into the tsunami generation process. Ocean Model. **56**, 43–56 (2012)
8. Kabanikhin, S.I.: Linear regularization of multidimensional inverse problems for hyperbolic equations (in Russian). Sobolev Institute of Mathematics. Preprint No. 27 (1988)
9. Romanov, V.G.: Inverse Problems of Mathematical Physics (in Russian). Nauka, Moscow (1984)
10. Romanov, V.G.: Stability in Inverse Problems (in Russian). Nauchniy Mir, Moscow (2005)
11. Kabanikhin, S.I., Krivorotko, O.I.: A numerical method for determining the amplitude of a wave edge in shallow water approximation. Appl. Comput. Math. **12**, 91–96 (2013)

12. Voronina, T.A., Tcheverda, V.A.: Reconstruction of tsunami initial form via level oscillation. Bull. Novosib. Comput. Cent., Ser. Math. Model. Geophys. **4**, 127–136 (1998)
13. Voronina, T.A.: Determination of spatial distribution of oscillation sources by remote measurements on a finite set of points. Siberian Journal of Numerical Mathematics **3**, 203–211 (2004)
14. Voronina, T.A.: Reconstruction of initial tsunami waveforms by a truncated SVD method. J. Inverse Ill-Posed Probl. **19**, 615–629 (2011)
15. Marchuk, A., Marinin, I., Komarov, V., Krivorotko, O., Karas, A., Khidasheli, D.: 3D GIS integrated natural and man-made hazards research and information system. In: Proceedings of The Joint International Conference on Human-Centered Computer Environments (HCCE) 2012, pp. 225–229. Aizu-Wakamatsu (2012)
16. Kabanikhin, S.I.: Inverse and Ill-Posed Problems: Theory and Applications. De Gruyter, Berlin (2011)

Hybrid Evolutionary Approach
to Multi-objective Mission Planning
for Group of Underwater Robots

Maksim Kenzin$^{(\boxtimes)}$, Igor Bychkov, and Nikolai Maksimkin

Matrosov Institute for System Dynamics and Control Theory,
Lermontov Str., 134, 664033 Irkutsk, Russia
gorthauers@gmail.com
http://www.idstu.irk.ru/

Abstract. We propose a hybrid approach, based on the combined use
of genetic algorithms, methods and heuristics of local search, and ant
colony optimization to solve the dynamic routing problem for a group
of underwater robots. Group's objective involves visiting a certain set of
control points (for the purpose of sampling, taking measurements, photos
and videos) according to their priority and under given restrictions. The
dynamic routing problem here is to find (*planning*) and adjust (*replanning*) feasible group routes for robots, ensuring as far as possible the
maximum efficiency of the group work.

Keywords: Autonomous underwater vehicles · Group control · Mission
planning · Transport routing problem · Evolutionary algorithms

1 Introduction

The rapid evolution of the subsea technologies in recent years has allowed underwater robots to take a significant role in the study of marine resources. These
autonomous underwater vehicles (AUVs) are widely used in such underwater
tasks as seabed mapping, mines searching and deactivating, oil and gas detection, taking samples, etc. At the moment, there is a clear tendency to increase
the degree of AUV's autonomy, allowing them to implement long-term underwater missions. Moreover, there is another tendency in robotics industry to switch
production of heavy-geared multitasking hand-built vehicles to serial assembly
of more reliable and cheap special purposed robots with modern onboard equip-
ment with extremely low power consumption. Application of coordinated and
distributed groups of such robots may significantly improve the effectiveness of
the large-scale underwater operations.

It is a problem of considerable practical interest to effectively route robots in
such works as surveillance and search, patrolling and inspection of underwater

The reported study was partly funded by RFBR according to the research projects
No. 14-07-00740-a, 14-07-31192-mol-a.

objects and structures, monitoring of water areas and topography mapping. All these tasks can be united by a common concept of a multi-objective mission within given area. One of the main features here are spatio-temporal constraints imposed by the specific nature of the water environment and by inaccuracy of the measuring devices.

In general, the multi-objective mission planning for the group of AUVs is a variation of vehicle routing problem (VRP) and consist of task assignment and path planning under given constraints. Depending on the type of research tasks and additional restrictions on the group movement, the routing problem may also obtain features of such VRP variations as routing with time windows, periodic routing and others.

All specialized VRP variations in scientific publications [1],[2] have one common feature remaining from the classical VRP: each control point (customer) must be visited exactly once (sometimes, not more than once) within one route. This restriction does allow to implement a search and surveillance missions [3], but can not be used for a long-term monitoring of dynamic processes and for other multi-objective missions that includes some sets of control points requiring periodic visits [4].

This paper proposes a formalized AUVs group routing problem for the periodic multi-objective missions and provides a hybrid evolutionary approach to address it effectively.

2   AUVs Mission Planning Problem

Periodic multi-objective mission requires AUVs to visit and examine the set of control points at scheduled intervals. The problem of planning such a mission is to find a feasible group route ensuring, as far as possible, the well-timed inspection of the majority of control points.

Group mission planning is carried out:

1. in a enclosed water area with a known seabed profile;
2. for a given finite set of control points (objectives) within the water area;
3. by a heterogeneous group of multiple AUVs;
4. under a certain set of restrictions imposed both on the group movement and movements of individual AUVs;
5. based on the effectiveness criteria of group work for current mission.

The mission area (Fig. 1a) is represented by three-dimensional space $D = \{(x, y, z) : 0 \leqslant x \leqslant X, 0 \leqslant y \leqslant Y, \zeta(x, y) \leqslant z \leqslant Z\}$, where function $\zeta(x, y)$ describes the seabed profile. *The mission area* may contain a set of forbidden zones, but anyhow, space D is always connected, which ensures the possibility of visiting of any point outside of restricted areas by AUVs.

Let Ω_N be a list of **mission objectives** (Fig. 1b) with their coordinates $(x_i, y_i, z_i) \in D$, $i = 1, ..., N$. Depending on the research task for each specific control point, the corresponding objective is assigned to one of two types

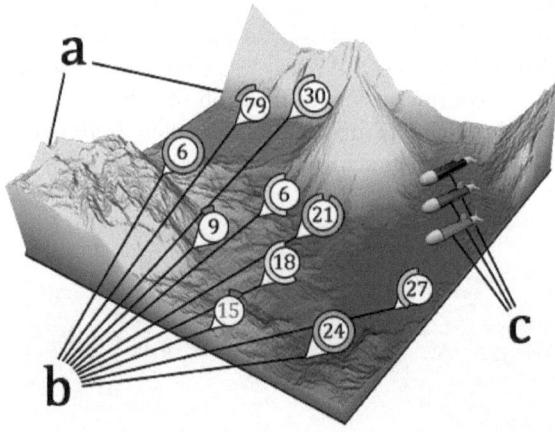

Fig. 1. Schematic representation of the periodic AUVs mission

$T_i = \{1, 2\}$ and receives its periodicity p_i. $T_i = 1$ means here that the periodicity of *objective* is *non-strict*, i.e. the duration of the time interval between its two successive inspections must not exceed p_i. The second *strictly periodic* type of objectives $T_i = 2$ requires exact length p_i of the described time period. In addition, each *objective* Ω_i requires some time s_i for its examination. Also let t'_i, $i = 1, ..., N$ be a time interval since the last inspection of each objective from Ω.

In other words, *non-strict objectives* should be inspected not less than once in their period p. The objectives of this type are commonly used in tasks, for which the principle "the more, the better" is appropriate: patrolling and guarding, checking the continuity of a different physical objects, etc. *Strict Objectives* should preferably be visited exactly at equal intervals. In that way, in case of arriving to the objective ahead of time, AUV should stand idle before starting the inspection. This procedure provides a more efficient way to study the dynamics of various underwater processes by AUVs taking samples and measurement, photo and video shoots.

The group of robots performing the mission consists of m functionally equivalent vehicles (Fig. 1c). At the same time, vehicles of the group may differ by their speed while moving between objectives. Let $(x^k, y^k, z^k) \in D$ be a current position of vehicle number k, and v^k be its speed. The speed of each robot allows us to calculate how long it will pass through specified segment of the path. In what follows we denote by c^k_{ij} the time required to k-th AUV to move from objective Ω_i to Ω_j, and by c^k_i the time to achieve objective Ω_i from AUV's current position.

The process of mission implementation is self-contained, i.e. all calculations are performed exclusively on board computer systems of robots. Group coordination is achieved by transfering data between members of the group through hydro-acoustic channel. Data can be transferred between vehicles both directly and through other AUVs. Thus, all vehicle must form a connected graph with lengths of arcs (distances between AUVs) less than the range of communication channel R_c to achieve complete synchronization of actual data within the group (Fig. 2). In what follows the group routes are called *communicatively stable*, if they guarantee the ability to synchronize data regularly.

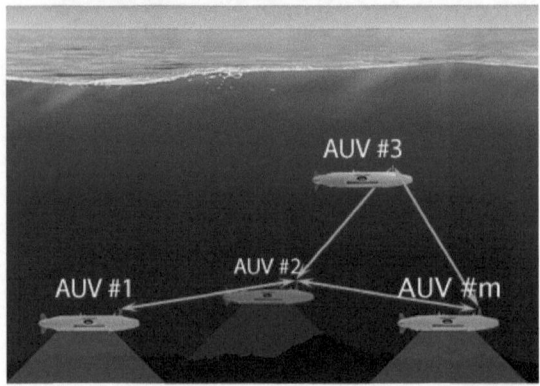

Fig. 2. Communication graph for the AUVs group

Communication stability requirement arises due to the dynamic nature of underwater missions: firstly, the observation results may require changing parameters of some objectives and even their removal; secondly, the uncertainty of external environment may lead to unexpected changes in the status of working group. All these changes may occur in real time, making it necessary to adjust the current route (replan) in order to maximize the group efficiency in new conditions. Among the events that require route replanning are:

- change of periodicity p_i and/or examination time s_i of an objective;
- change of type T_i of an objective;
- adding new objectives or removing existing ones;
- change of the AUVs group's composition.

The route of single vehicle $r = \langle v_1(r), v_2(r), ..., v_h(r) \rangle$ is a list of objective's numbers in the consecutive order of their planning visit. It should be noted that any objective can be included more than once into the route of a single robot. Group route $G = \{r_1, ..., r_m\}$ is a set of single AUV routes.

One of the main characteristics of the route is its time duration $t(r)$:

$$t(r) = c^j_{v_1(r)} + o_{v_1(r)} + s_{v_1(r)} + \sum_{i=2}^{h}(c^j_{v_i(r)v_j(r)} + o_{v_i(r)} + s_{v_i(r)}), \qquad (1)$$

where o_i is an idle time before starting examination of the i-th objective. If the *objective is non-strict* ($T_i = 1$) then $o_i = 0$, in another case ($T_i = 2$) the idle time is defined as $o_i = max\{p_i - t'_i, 0\}$.

The effectiveness of the group work is determined, first of all, by regularity of scheduled inspections. Situations, when AUV arrives too late $t'_i > p_i$ and delays inspection of the objective, are undesirable and should be penalized via efficiency criteria. If it is physically impossible to visit all objectives in time, the criteria should consider two possible types of solutions:

- routes that provide minimal delay time with all mission objectives are being visited;
- routes that guarantee full absence of delays by ignoring some objectives.

Thus, the **routing problem** is to find a feasible group route that provides the minimum delay time.

2.1 Main Features of the Problem

To solve the problem described above the following features has to be considered in the first place:

- undefined duration of the whole mission;
- dynamic conditions of the mission;
- communication stability requirement;
- expectable large-size of the problem.

Given these features we suggest the following decomposition approach: to divide the process of mission implementation on a finite time periods (*periods of planning*) with data synchronization within the group at the end of each period. In this case, the process of mission implementation on a single period of planning can be represented as follows on the block scheme (Fig. 3).

Accordant to this scheme each vehicle computes to find the best group route for the next period of movement while following their pre-planned routes on the current one. Full data synchronization at the end of each period allows robots to receive all information obtained by the group and to exchange their best found solutions. Thus, the most effective route among suggested is selected to become approved route for the new planning period. This approach allows to parallelize all calculations among the robots in a natural way.

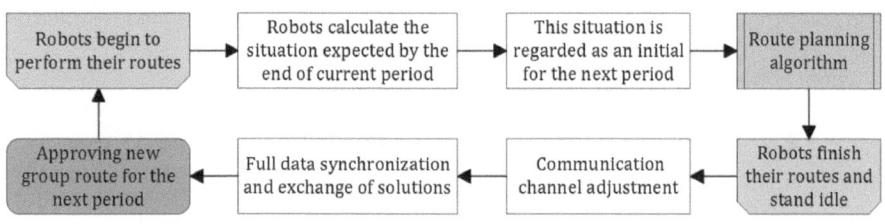

Fig. 3. AUVs group functioning during a single period of planning

In case of route planning for a single limited period the initial routing problem would be a variation of the VRP, which is known to be NP-hard. The main features of the routing problem for a single period are as follows:

1. there is no depot point to arrive at the end of period, thus the final position of all vehicles has to be assured to allow data synchronization;
2. the expected mission condition at the end of current period should also be taken into account, since it will be the initial condition for the next period;
3. the necessity to choose objectives which inspection should be delayed or canceled if it is impossible to visit all mission objectives in time.

2.2   Efficiency Criteria for the Group Route

Limited duration of planning period allows us to construct an efficiency criteria for the group route in an explicit form. In order to do this, we define an additional function $a_i(t)$ corresponding to each objective Ω_i of the mission which we call the *function of relevance*, and its value at the moment t - the *relevance of the objective*:

$$a_i(t) = \begin{cases} \delta_i(t - (t_{ik} + s_i)), & t \in [t_{ik} + s_i, t_{ik+1}) \\ 0, & t \in [t_{ik}, t_{ik} + s_i] \end{cases}, k = 1, 2, ... \tag{2}$$

where t_{i1}, t_{i2} is a sequence of moments, when i-th objective is expected to be visited by AUVs according to the current route, and δ_i is non-decreasing *function of the relevance growth*, $\delta(0) = 0$. The examination of an objective Ω_i by a robot resets the relevance of the objective to a *zero* value, following that the function a_i begins to increase and reaches its value $\bar{a}$ (equal for all objectives) in time period of p_i.

The function of the relevance growth δ_i determines both the extent of the need for visiting corresponding objective and penalty for delaying its inspection. We suggest using the simple linear function $\delta_i = \bar{a}t/p_i$ (Fig. 4), which is not only easy to implement, but also assigns a relative importance of all the objectives in a way to ensure well-timed inspections of the objectives with the least periodicity.

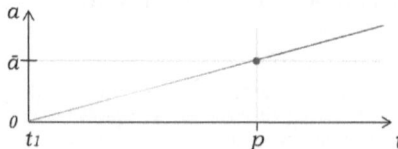

Fig. 4. Objective relevance increases after its last visit by an AUV

Now we define a function for calculating a penalty for delaying the inspection of the objective:

$$\varphi(a_i, t) = \begin{cases} a_i(t) - \bar{a}, & a_i(t) > \bar{a} \\ 0, & a_i(t) \le \bar{a} \end{cases}. \tag{3}$$

Hence, the total penalty for the group route G is:

$$\Phi(G) = \sum_{i=1}^{m} \sum_{j=2}^{h} \varphi(a_v, t_v), \tag{4}$$

where v stands for $v_j(r_i)$ and the t_v values represent the time moments of their expected inspections. The sequence of moments t_v for each objective is calculated through a simulation procedure. Operating with critera (4) only will lead us to the situations, when some objectives are being ignored to exclude all delays during group movement. Thus, it is needed to consider an additional function, which would "care" not about objectives within a group route as (4), but about the whole sitation by the end of period:

$$\Psi(G) = \sum_{i=1}^{N} a_i(t(G)), \tag{5}$$

where $t(G) = \max\limits_{j=1,\dots,m} (t(r_j))$ - the time moment of the route completion. The function (5) delivers a some sort of forecasting value for the next period of planning. Using (5) as a part of criteria allows us to provide group routes with two positive features: it forbids robots to ignore the objectives of the mission and also indirectly normalises route durations of all vehicles in the group.

Hence, the final efficiency criteria for the group route is as follows:

$$f(G) = \omega \cdot \Phi(G) + \Psi(G), \ G \in Z, \tag{6}$$

where Z is the set of all possible communicatively stable routes and ω is the weight number that allows one to expertly choose the operating mode of the group in case of impossibility of well-timed inspection of all mission objectives: with $\omega \cong 1$ the group will always examine all the mission objectives but with some delays (Fig. 5 left); with $\omega \gg 1$ the group will continue to inspect only those objectives which can be visited strictly in time, and the remaining objectives will be ignored (Fig. 5 right).

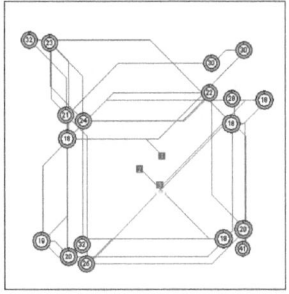 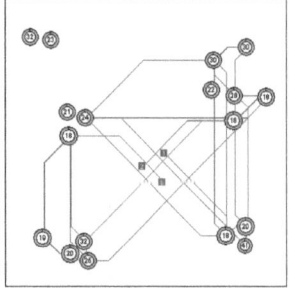

Fig. 5. Two different operating modes of the group

3   Hybrid Evolutionary Approach to AUVs Mission Planning

For a broad class of VRP there are no algorithms solving it in polynomial time, which leads us to the class of approximation algorithms, that allow to obtain rational sub-optimal solutions in low computational time. The routing problems can be solved by significant diversity of methods of operations research but there is no unified algorithm for serving the full spectrum of problems of this class.

A comprehensive survey of various approaches for solving the VRP concludes that evolutionary algorithms (EAs) generally outperform any other heuristic or metaheuristic [5]. Its main advantage is ability to find solutions for poorly structured problems and problems with complex constraints, as EAs require a relatively small amount of information about the nature of the problem. Furthermore, the efficiency of EAs can be significantly increased by combining with a local search and improvement heuristics.

The described above AUV routing problem has a "bad" neighborhood structure, making it difficult to allocate and find qualitative and feasible solutions, as they may not be in the neighbourhood of another feasible solutions of high quality in the search space. Evolutionary algorithm for solving the given problem must possess specialized genetic operators featuring local search heuristics and other methods of local solutions improving. The block diagram for the proposed evolutionary hybrid algorithm is shown in Fig. 6.

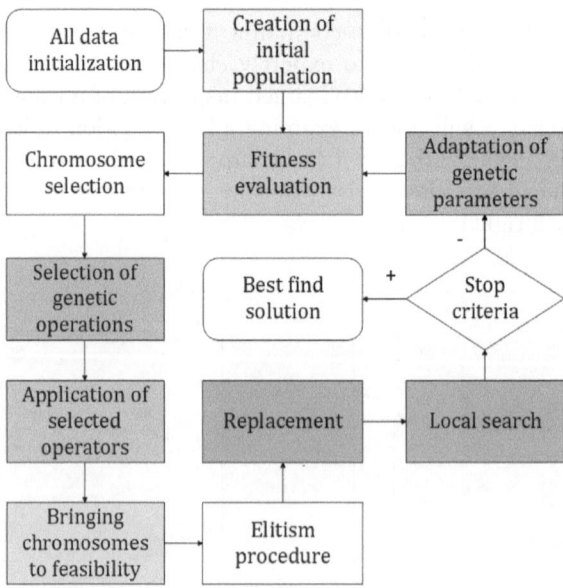

Fig. 6. The block diagram of hybrid evolutionary approach with colored blocks of author's procedures and modifications

The algorithm starts with initialization of all data including the latest list of objectives, the current state of the group and all algorithm parameters. Then, the initial population of chromosomes (group routes) is created to ensure both covering a significant portion of the search space and containing a variety of good solutions. This requirement is achieved by the simultaneous use of three different construction heuristics: a sequential insertions and two parallel insertions.

The generated population is evaluated by an objective function (6) that takes into account not only the effectiveness of the route during current planning period, but also an estimation of the search complexity on the next scheduling period. According to the results of ranking, the tournament selection chooses a set of chromosomes to be used for procreation and/or mutations.

The basic EA proposes the scheme with a chromosome supposed firstly to be crossed with another candidate solution and after to become a subject of mutation. We suggest to modify such a scheme, because it can sometimes lead to loss of the intermediate genetic information, and thus to expose each solution either to only one genetic operator or to both of them in a common way.

Specialized genetic operators here consist of multimode mutation and two different variants of crossover. The first crossover is a heuristic modification of the two-point crossover; the second crossover operator seeks to combine the general characteristics of the parental individuals with the qualities of the best known solution.

The mechanisms of parallel populations with immigration (island model), clone removal and elitism provide faster algorithm convergence rate while preventing premature convergence to local optima.

In order to further improve our newly received individuals the deterministic heuristic search techniques of *2-opt exchange* and *λ-interchange* are applied periodically to the whole population for a deeper exploitation.

The evolutionary process is repeated until a stop criteria has been met. Here such a criteria is an impossibility to improve solutions with any of the presented local search heuristics.

The Procedure to Bring Solutions to Feasibility. To ensure overall *communication stability* of the cumulative group route, vehicles of the group should be able to synchronize data at the end of each period. To achieve this condition, each constructed route is iteratively modified according to the following procedure until requirements are met:

1. The graph formed by final positions of all robots $v_h(r_k)$, $k = 1, ..., m$ according to their routes is checked for connectivity only on those arcs whose length does not exceed the range of communication channel R_c.
2. If the graph is disconnected, the center of gravity $g \in D$ of all AUVs final positions is calculated.
3. The most distant from g objective is defined between all final objectives of single vehicle routes and replaced in the corresponding route by another objective, that is closest to g.
4. Return to Step 1.

Genetic Parameters Adaptation Mechanism. To improve the efficiency of a population creation procedure, all probabilistic genetic parameters are changed constantly at the end of each iteration of the algorithm. Being determined by the current efficiency of the genetic operators these changes allow algorithm to adapt to the different situations on different steps of processing. The implementation of the adaptation mechanism may significantly increase the speed of computing in those cases when some genetic operators begin to work significantly better than others. The adjustable genetic parameters here are (Fig. 7):

- the probability of choosing either Crossover $P(C)$ or Mutation $P(M)$ for each chromosome;
- the probability of each crossover operator C_i;
- the probability of each mutation mode M_j;
- the probability P_{ij} to choose a mutation mode M_j after crossover C_i.

In general, the adaptation mechanism is to keep track of the paths, which lead to the solution improving more often than others, and to increase their probabilities for the next generation of solutions. The scheme of the ant colony optimization algorithm allows to implement such an adaptation in a natural way.

In our case, each chromosome of the evolutionary algorithm represents both a solution and an ant, which goes through a network of genetic operators and marks them with pheromones, if their work has improved the individual itself. At the end of each iteration of the algorithm the probabilities for all operators are redistributed according to the amount of pheromone. With the pheromones evaporation mechanism, the initial distribution of probabilities (e.g. equally probable) would be restored eventually if the current set of genetic parameters would lose its relevance.

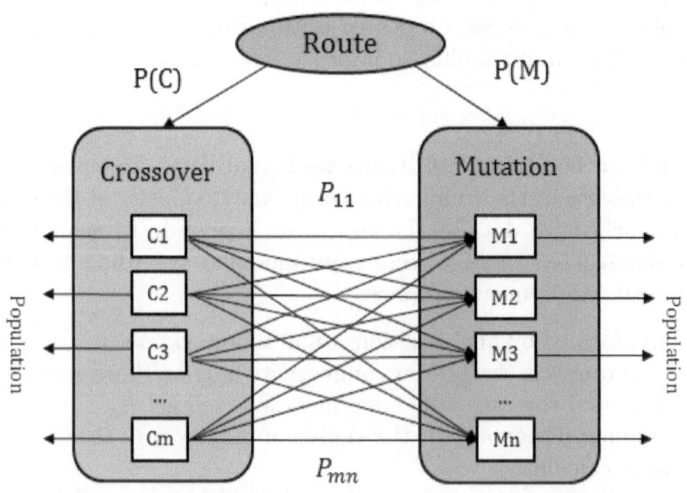

Fig. 7. The general scheme of genetic operators network

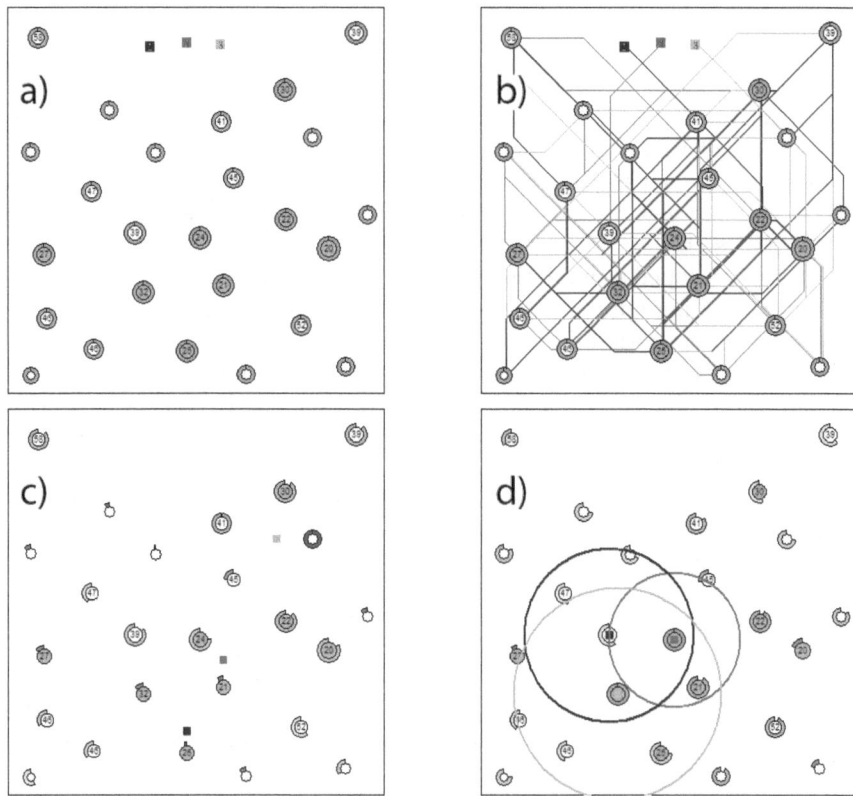

Fig. 8. Example of the mission planning for the group of 3 AUVs

4   The Results of Test Calculations

On the basis of a series of simulation experiments using the developed framework AUV Mission Planner the best initial parameters of the algorithm were found to guarantee its efficiency on various types of test problems. As a result, the average deviation of the generated solutions from the optimal ones was obtained.

   In the following example (Fig. 8), we demonstrate how the algorithm manages to generate the group route for 3 AUVs in the mission with 25 objectives (Fig. 8a) of both types of regularity (white and grey). The periodicity values p_i of the objectives lie in the interval between 1000 to 5000 time units, the whole planning period is 12000 time units (more than 50 scheduled visits of objectives). Within computing time of 3 minutes the algorithm calculates the group route (Fig. 8b) providing well-timed inspection of the majority of objectives with a total delay of 101 time units while optimal solution provides 37 (deviation of $\sim$0,53%). Figure 8c shows a screenshot of AUV Mission Planner during the simulation of the group movement. The route (Fig. 8b) is built to ensure the possibility of data synchronization within the group at the end of the planning period (Fig. 8d).

5   Conclusion

The routing problem statement proposed in the paper allows us to formalize a multi-objective mission of regular monitoring and can be distinguished as a separate subclass of VRP, which combines features of the PVRP and the VRPTW. The problem is open to be expanded with new entities, parameters and constraints, such as new various types of objectives, functionally heterogeneous group of AUVs etc.

We have developed a simulation framework AUV Mission Planner to test and work out scheduling approaches and mission planning algorithms for groups of autonomous robots with different types of tasks, objectives and restrictions. The high efficiency of the suggested algorithm is shown through simulation studies in solving the problems of scheduling group missions under various conditions. The algorithm is proved to be used for efficient multiscale routing of automated robots as a high-level control algorithm. At the same time, the hybrid approach of described structure, as well as the proposed set of heuristics and procedures, can also be applied for solving both standard vehicle routing problem and its various modifications.

References

1. Chow, B., Clark, C.M., Huissoon, J.P.: Assigning closely spaced targets to multiple autonomous underwater vehicles. Journal of Ocean Technology **6**(1), 46–68 (2011)
2. Deng, Y., Beaujean, P.-P.J., An, E., Carlson, E.: Task allocation and path planning for collaborative autonomous underwater vehicles operating through an underwater acoustic network. Journal of Robotics **2013**, 1–15 (2013)
3. Kiselyov, L.V., Inzartsev, A.V., Bychkov, I.V., Maksimkin, N.N., Hmelnov, A.E., Kenzin, M.Yu.: Situational control by group of autonomous underwater robots on the basis of genetic algorithms. Underwater Investigation and Robotics **2**(8), 34–43 (2009). (In Russian)
4. Bychkov, I.V., Kenzin, M.Yu., Maksimkin, N.N., Kiselev, L.V.: Evolutionary approach to group routing of autonomous underwater vehicles in dynamic multiobjective monitoring missions. Underwater Investigation and Robotics **2**(18), 4–13 (2014). (In Russian)
5. Braysy, O., Gendreau, M.: Vehicle routing problem with time windows, Part I: Route construction and local search algorithms. Transportation Science **39**(1), 104–118 (2005)

Theoretical and Numerical Prediction of the Permeability of Fibrous Porous Media

Aziz Kudaikulov[1]([⊠]), Christophe Josserand[2], and Aidarkhan Kaltayev[1]([⊠])

[1] Al-Farabi Kazakh National University, Almaty, Kazakhstan
aziz.kudaikulov@gmail.com,
aidarkhan.kaltayev@kaznu.kz
[2] Sorbonne Universités, Institut D'Alembert, CNRS and UPMC UMR 7190,
4 place Jussieu, 75005 Paris, France
christophe.josserand@gmail.com

Abstract. In this paper, the permeability of ordered fibrous porous media for normal flows is predicted theoretically and numerically. Moreover, microscopic velocity profiles in the "unit cell" are investigated in detail for normal flows. Porous material is represented by a "unit cell" which is assumed to be repeated throughout the media and 1D fibers are modeled. Fibers are presented as cylinders with the same radii. Planar flow that perpendicular to the axes of cylinders is considered in this paper. All numerical calculations are performed using Gerris program [6]. The quantitative comparison of numerical and theoretical results of computation of the permeability of ordered fibrous media is reasonably good and is about 10–15%.

Keywords: Fibrous porous media with periodic structure · Navier-stokes equations · Darcys law · Permeability of fibrous porous media

1 Introduction

Fibrous porous materials are widely used in modern industry and engineering applications, such as heat exchangers, filters, catalysts, and fuel cell electrodes. The main technical challenge for the fibrous porous medium is to determine the velocity of the flow in the media. If we know the velocity of the fluid flow in the fibrous porous media, we can determine the important technical features of the media, such as the rate of change of temperature of the medium, the rate of change of concentration of substance, etc.. In most cases, the flow in the fibrous porous media is very slow and obeys Darcy's law [1], that relates the flow rate to the pressure gradient:

$$\boldsymbol{u}_d = \frac{K}{\mu}\nabla(p + \rho g z), \tag{1}$$

where K - permeability of fibrous porous medium, μ - fluid viscosity, $\boldsymbol{u}_d$ - flow rate, p - pressure in the porous medium and $\rho g z$ - hydrostatic pressure. Calculating

© Springer International Publishing Switzerland 2015
N. Danaev et al. (Eds.): CITech 2015, CCIS 549, pp. 85–93, 2015.
DOI: 10.1007/978-3-319-25058-8_9

of the flow rate from the formula (1) is very difficult problem, because we don't know the permeability of fibrous porous medium generally. Basically permeability of the porous medium K is determined empirically, however new experimental technologies and high-resolution imaging of porous media can provide three-dimensional structural details of porous materials with resolution in one micron. There exist many pore-scale models of the fluid flows in the porous medium such as Lattice-Boltzmann [3], pore network models [2], discrete particle methods (smoothed particle hydrodynamics) [4] and direct discretization methods (finite difference, finite element, finite volume, immersed boundary methods) [5]. All of these methods require high computational power. In the direct discretization methods the Navier-Stokes equations are discretized and solved for domain with complex geometries. The main advantages of the direct discretization methods is that can be applied for the domains with complex geometries and simulate fluid flow in the porous medium more exactly than others. Also these methods have disadvantages, such as: it requires high computational power and these methods are available for very small domains (about 1 micron). From the above listed models, the most preferred is pore network models [2]. In this model, the porous medium is represented as a system of straight channels. Nevertheless, in many engineering designs the geometry of the porous medium is very simple. Also, in many engineering calculations no need to accurately calculate the parameters of fluid flow in porous media, but it is sufficient to calculate their average values. The fibrous porous medium with a periodic structure is considered in this paper (see Fig. 1). Fibers are presented as cylinders with the same radii. Planar flow that perpendicular to the axes of cylinders is considered in this paper. We numerically predicted the permeability of fibrous porous media and compared with existing theoretical predictions in this study. All numerical calculations are performed using Gerris program [6].

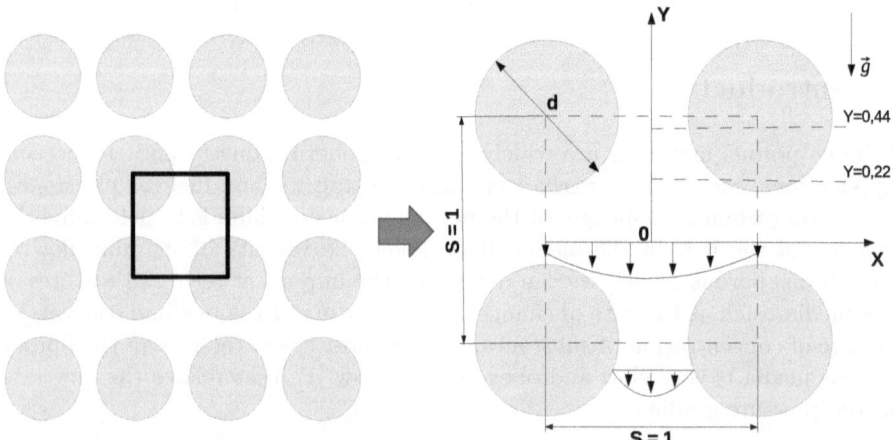

Fig. 1. Two dimensional rectangular area with sizes s x s in that the cylinders are periodically arranged

2 Modeling Approach

There exist many theoretical predictions of the permeability of fibrous porous media in the literatures [7] - [10]. From early works on the theoretical predictions of the permeability of fibrous porous medium we can emphasize the works of John Happel (1959) [7] and Hasimoto (1959) [8]. John Happel [7] found the theoretical prediction of the permeability of fibrous porous media by solving the Stokes equation for a fluid flow in fibrous porous medium. The flow around a cylinder investigated in his work (see Fig. 2). His theoretical prediction of the permeability of fibrous porous medium:

$$K_1^* = \frac{K_1}{d^2} = \frac{1}{32\phi}[ln(\frac{1}{\phi}) - \frac{1-\phi^2}{1+\phi^2}], \qquad (2)$$

where K_1 - permeability of fibrous porous media, d - diameter of the cylinders and $\phi = \frac{d^2}{s^2}$, where s - the distance between centers of the cylinders.

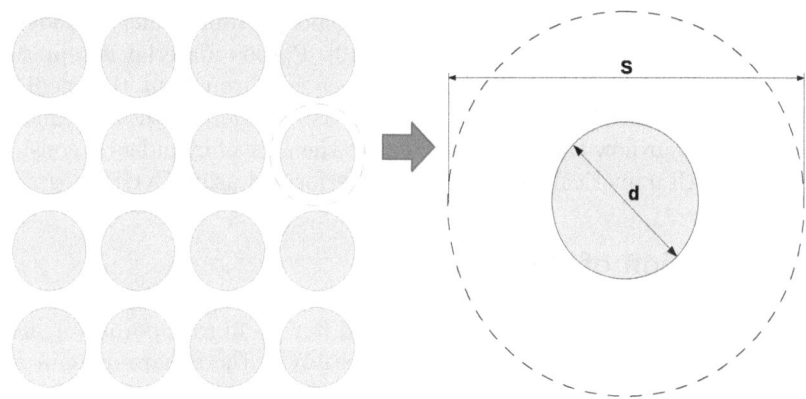

Fig. 2. The periodic structure of the fibrous porous medium

In the work of Hasimoto [8] the exact solution of the Stokes equation for the fluid flow in fibrous porous medium in the form of the infinite series is used to predict the permeability of fibrous porous media. He found the theoretical prediction of the permeability of fibrous porous media using only the terms of lowest order of this series:

$$K_2^* = \frac{K_2}{d^2} = \frac{1}{32\phi'}[ln(\frac{1}{\phi'}) - 1,476], \qquad (3)$$

where $\phi' = \frac{\pi d^2}{4s^2}$. Later Sangani and Acrivos (1982) [9] improved the theoretical prediction of fibrous porous media using the terms of highest order of the series that presented in the work of Hasimoto [8]:

$$K_3^* = \frac{K_3}{d^2} = \frac{1}{32\phi'}[ln(\frac{1}{\phi'}) - 1,476 + 2\phi' - 1,774\phi'^2 + 4,076\phi'^3]. \qquad (4)$$

From recent works on the theoretical predictions of the permeability of fibrous porous medium we can emphasize the work of Tamayol and Bahrami (2008) [10]. In this work, porous medium is considered as "unit cell" which is repeated throughout the media (see Fig. 1). Also the unidirectional flow with parabolic velocity profile is considered in this work. Their theoretical prediction of the permeability of fibrous porous medium:

$$K_4^* = \frac{K_4}{d^2} = \frac{1}{3\phi^2} \frac{(1-\phi)^{\frac{5}{2}}}{(2(\phi+2) + 4\frac{(1-\sqrt{\phi})(1-\phi)^2}{\sqrt{\phi}})\frac{\sqrt{1-\phi}}{\sqrt{\phi}} + 12arctan(\frac{1+\sqrt{\phi}}{\sqrt{1-\phi}})}. \qquad (5)$$

In this paper, the permeability of ordered fibrous porous media for normal flows is calculated numerically and compared with the above theoretical predictions. Moreover, microscopic velocity profiles in the "unit cell" are numerically calculated and compared with the parabolic velocity profile which considered in the work of Tamayol and Bahrami (2008) [10]. Porous material is represented by a "unit cell" which is assumed to be repeated throughout the media and 1D fibers are modeled. Fibers are presented as cylinders with the same radii (see Fig. 1). Planar flow that perpendicular to the axes of cylinders is considered in this paper. All numerical calculations are performed using Gerris program [6].

3   Formulation of the Problem

The numerical simulation of single-phase fluid flow in fibrous porous medium is considered in this paper. The object of the study is the square domain which includes 4 cylinders (see Fig. 1). Planar flow that perpendicular to the axes of cylinders is considered in this paper. This model is based on the numerical solution of the Navier-Stokes equations for incompressible fluid flow:

$$\frac{\partial u}{\partial t} + (u \cdot \nabla)u = g - \nabla p + \frac{1}{Re}\nabla^2 u, \qquad (6)$$

$$\nabla \cdot u = 0, \qquad (7)$$

where u - velocity of fluid flow, p - pressure and g - acceleration of gravity.

Initial condition for the velocity of fluid flow:

$$u(0, x_k) = 0. \qquad (8)$$

Boundary conditions for the velocity and pressure:

1) On the boundary of the domain (on the Fig. 1 showed as dashed lines):

$$\boldsymbol{u}(t,x_k)|_{x_k=-s/2} = \boldsymbol{u}(t,x_k)|_{x_k=s/2}, \tag{9}$$

$$p(t,x_k)|_{x_k=-s/2} = p(t,x_k)|_{x_k=s/2}, \tag{10}$$

$$\frac{\partial \boldsymbol{u}(t,x_k)}{\partial x_k}|_{x_k=-s/2} = \frac{\partial \boldsymbol{u}(t,x_k)}{\partial x_k}|_{x_k=s/2}. \tag{11}$$

2) On the surface of the cylinders (no-slip boundary condition):

$$\boldsymbol{u}(t,x_k) = 0, \tag{12}$$

where $k = 1,2$ (for two-dimensional case). We need to average the value of the velocity over the square domain (on the Fig. 1 showed as dashed lines) to find the flow rate - $\boldsymbol{u}_d$:

$$\boldsymbol{u}_d = \frac{\int\int \boldsymbol{u}(t,x)dxdy}{s^2}, \tag{13}$$

where s - the distance between centers of the cylinders. Since we only considered the flow, when $Re \ll 1$, then we can neglect the horizontal component of the velocity in comparison with the vertical component of the velocity and due to the incompressibility of the fluid we have:

$$\int U(t,x)dy + \int V(t,x)dx \approx \int V(t,x)dx = const, \tag{14}$$

where $U(t,x)$ - the horizontal component of the velocity, and $V(t,x)$ - the vertical component of the velocity:

$$\int V(t,x)dx = Q = const, \tag{15}$$

where Q - flow rate. From the equations (13) and (15) follows:

$$V_d = \frac{\int\int V(t,x)dxdy}{s^2} = \frac{\int Q dy}{s^2} = \frac{Q}{s}. \tag{16}$$

Further, from the equation (1) we can find the permeability of the porous medium:

$$K = |\frac{\mu \boldsymbol{u}_d}{\nabla(p+\rho gz)}| = \frac{\mu Q}{s\nabla(p+\rho gz)}. \tag{17}$$

4   Results

On the Fig. 3, 4 and 5 shows the comparison of the numerical and theoretical velocity profile which given in the work of Tamayol and Bahrami (2008) [10] for various values of the radius of the cylinders. The parabolic velocity profile is considered in the work of Tamayol and Bahrami (2008) (see Fig. 1):

$$V(x,y) = ax^2 + bx + c. \tag{18}$$

Since the fluid flow is incompressible, we have:

$$V(x,y) = -\frac{3Q}{4\delta^3}(x^2 - \delta^2) = \begin{cases} -\frac{3Q}{4(\frac{s}{2}-\sqrt{\frac{d^2}{4}-y^2})^3}(x^2 - (\frac{s}{2} - \sqrt{\frac{d^2}{4} - y^2})^2), 0 \le x \le \frac{d}{2} \\ -\frac{3Q}{2s^3}(4x^2 - s^2), \frac{d}{2} \le x \le \frac{s}{2}. \end{cases} \tag{19}$$

where δ - distance between the surfaces of the cylinders. Also comparison of the numerical value of the permeability of fibrous porous media with above theoretical predictions is showed on the Fig. 6 and Table 1.

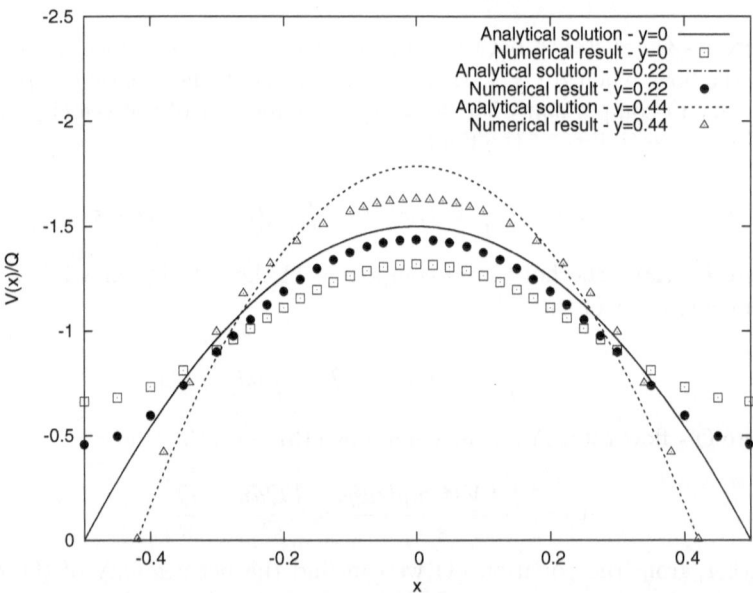

Fig. 3. The numerical and theoretical profile of the vertical component of the velocity when the diameter of the cylinders - d = 0,2

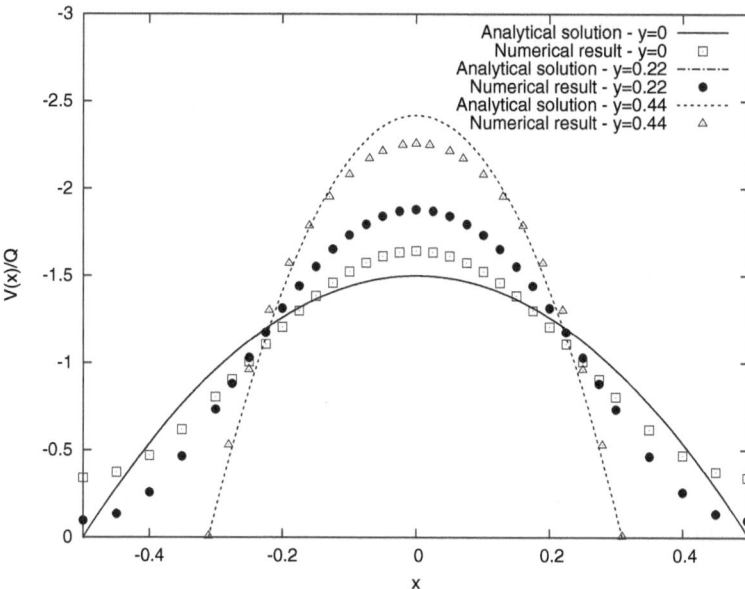

Fig. 4. The numerical and theoretical profile of the vertical component of the velocity when the diameter of the cylinders - d = 0,4

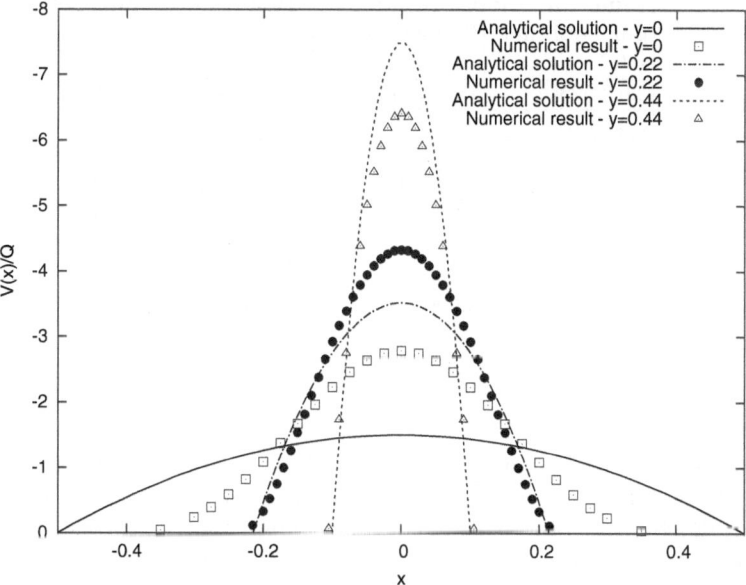

Fig. 5. The numerical and theoretical profile of the vertical component of the velocity when the diameter of the cylinders - d = 0,8

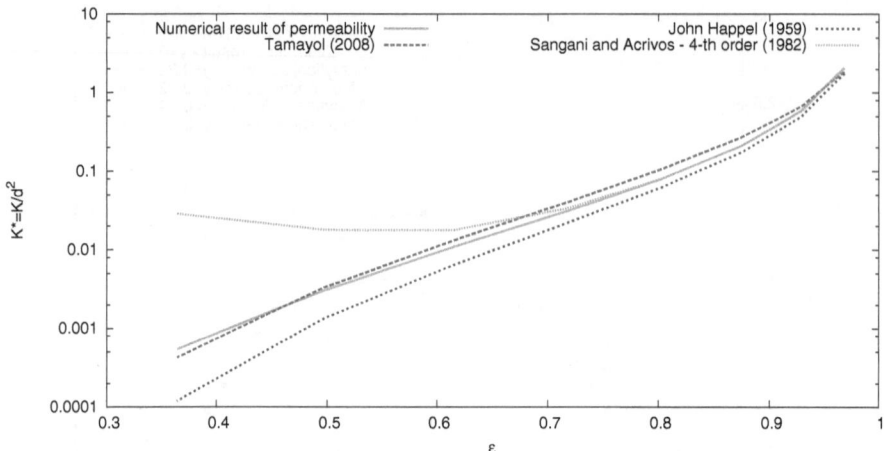

Fig. 6. A comparison of numerical and theoretical values of the permeability of fibrous porous medium

Table 1. A comparison of numerical and theoretical values of the permeability of fibrous porous medium

Porosity - ϵ	Numerical value of the permeability - K^*	Tamayol and Bahrami (2008) - K_4^*	John Happel (1959) - K_1^*	Sangani and Acrivos (1982) - K_3^*
0.9686	2.076525	1.828917	1.735993	2.036350
0.9294	0.584678	0.661806	0.494450	0.578512
0.8744	0.207850	0.267379	0.172364	0.206495
0.8037	0.080000	0.107103	0.062993	0.080788
0.7174	0.030931	0.040066	0.021798	0.033532
0.6154	0.010859	0.013148	0.006414	0.017729
0.4976	0.003048	0.003323	0.001340	0.017806
0.3641	0.000543	0.000424	0.000118	0.028461

5   Conclusions

As can be seen from Fig. 6 and Table 1, the theoretical prediction of the permeability of the fibrous porous medium which given in the work of Tamayol and Bahrami (2008) [10] is the most accurate in comparison with other theoretical predictions. Also in Fig. 3, 4 and 5 showed that the error of the theoretical prediction of the velocity profile is not so large and allows to investigate in detail the fluid flow in pore scale or micro scale.

References

1. Bear, J., Cheng, A.H.-D.: Theory and Applications of Transport in Porous Media. Modeling Groundwater Flow and Contaminant Transport, vol. 23. Springer (2010)
2. Blunt, M., King, P.: Relative permeabilities from two- and three-dimensional pore-scale network modelling. Transport in Porous Media **6**(4), 407–433 (1991)
3. Pan, C., Hilpert, M., Miller, C.T.: Lattice-Boltzmann simulation of two-phase flow in porous media. Water Resources Research **40**(1), W01501 (2004)
4. Tartakovsky, A.M., Meakin, P.: Pore scale modeling of immiscible and miscible fluid flows using smoothed particle hydrodynamics. Advances in Water Resources **29**(10), 1464–1478 (2006)
5. Huang, H., Meakin, P., Liu, M.B.: Computer simulation of two-phase immiscible fluid motion in unsaturated complex fractures using a volume of fluid method. Water Resources Research **41**(12), W12413 (2005)
6. Popinet, S.: The Gerris Flow Solver. http://gfs.sourceforge.net
7. Happel, J.: Viscous flow relative to arrays of cylinders. AIChE **5**, 174–177 (1959)
8. Hasimoto, H.: On the periodic fundamental solutions of the Stokes equations and their application to viscous flow past a cubic array of spheres. J. Fluid Mech. **5**, 317–328 (1959)
9. Sangani, A.S., Acrivos, A.: Slow flow past periodic arrays of cylinders with application to heat transfer. Int. J. Multiphase Flow **8**, 193–206 (1982)
10. Tamayol, A., Bahrami, M.: Analytical determination of viscous permeability of fibrous porous media. International Journal of Heat and Mass Transfer **52**, 2407–2414 (2009)

A Study of (m,k)-Methods for Solving Differential-Algebraic Systems of Index 1

Alexander I. Levykin[1] and Eugeny A. Novikov[2](✉)

[1] Institute of Computational Mathematics and Mathematical Geophysics, Academy of Sciences, Siberian Branch, pr. Ak. Lavrent'eva 6, Akademgorodok, 630090 Novosibirsk, Russia
lai@osmf.sscc.ru
http://www.sscc.ru/
[2] Institute of Computational Modeling, Academy of Sciences, Siberian Branch, Akademgorodok 50, Str. 4, 660036 Krasnoyarsk, Russia
novikov@icm.krasn.ru
http://icm.krasn.ru/

Abstract. A class (m,k)-methods is discussed for the numerical solution of the initial value problems for implicit systems of ordinary differential equations. The order conditions and convergence of the numerical solution in the case of implementation of the scheme with the time-lagging of matrices derivatives for systems of index 1 are obtained. At $k \leq 4$ the order conditions are studied and schemes optimal computing costs are obtained.

Keywords: Stiff systems · Differential-algebraic systems of index 1 · Numerical methods

1 Introduction

Many applied problems lead to systems of differential equations given implicitly as [1–4]

$$F(x, x') = 0, \ x(t_0) = x_0, \ t_0 \leq t \leq t_k, \tag{1}$$

where x and F are functions of the same dimension, and F is assumed to have sufficiently many bounded derivatives. Such problems arise in simulation of chemical reaction kinetics [4], electrical networks [5–6], control engineering etc. A non-autonomous systems $F(x, x', t) = 0$ is brought to the form (1) by adding the equation for the independent variable, $t' = 1$.

The modern methods for numerical solution of the initial-value problem for systems of ordinary differential equations (ODE) suppose usually the explicit dependence of the derivative of the solution [7]

$$x' = \varphi(x), \ x(t_0) = x_0, \ t_0 \leq t \leq t_k. \tag{2}$$

Support of RFBR Ander Grants 14-01-00047 and 15-01-00977.

N. Danaev et al. (Eds.): CITech 2015, CCIS 549, pp. 94–107, 2015.
DOI: 10.1007/978-3-319-25058-8_10

However, a reduction of (1) to the form (2) requires a large additional numerical costs at every integration step, because this is connected with the inversion of the matrix $F_y = \partial F / \partial y$ which generally is singular. The numerical problem appeares to be more complicated because of the stiffness of explicit equations systems: in this case it is necessary to apply of special methods with conversion of the Jacobian matrix . A class of the schemes is offered [8], in which the reduction to the form (1) and the calculation of the approximate solution are carried out simultaneously. The given methods were generated by the (m, k)-schemes [9] for solving the explicit ODE systems.

We use classification of implicit systems, based on the concept of the index for such systems [1–2]. We say that system (1) is:

a) of index 0, if $\|F_y^{-1}\| \le c < \infty$ (i.e., when (1) is solvable);

b) of index 1, if (1) can be reduced to

$$x' = f(x, y), \ 0 = g(x, y), \tag{3}$$

where $\|g_y^{-1}\| \le c < \infty$;

c) of index 2, if (1) can be reduced to

$$x' = f(x, y), \ 0 = g(x),$$

where $\|(g_x f_y)^{-1}\| \le c < \infty$.

In addition, it is assumed that functions F, f, and g are Lipschitz bounded, which ensures existence and uniqueness of the solution to problem (1) [10].

Using the notation $x' = y$, problem (1) can be written in the form

$$x' = y, \ F(x, y) = 0, \ x(t_0) = x_0, \ y(t_0) = y_0, \ t_0 \le t \le t_k. \tag{4}$$

The additional condition $y(t_0) = y_0$ can be found, for example, by solving the problem $F(x_0, y) = 0$ and using the stabilization technique.

2   The Numerical Schemes

We define the class of the (m, k)–schemes for solving problem (4). Let m and k, $(m \ge k)$ be given integers and consider the sets

$$\begin{aligned} M_m &= \{1, \ldots, m\}, \\ M_k &= \{m_i \,|\, m_1 = 1, \ m_{i-1} < m_i, \ m_i \le m, \ 2 \le i \le k\}, \\ J_i &= \{m_j - 1 \,|\, j > 1, \ m_j \in M_k, m_j \le i\}, \quad 2 \le i \le m. \end{aligned} \tag{5}$$

Then (m, k)-methods for the systems of index 0 have the form

$$x_{n+1} = x_n + \sum_{i=1}^{m} \mu_i k_{xi}, \quad y_{n+1} = y_n + \sum_{i=1}^{m} \mu_i k_{yi}, \tag{6}$$

where the internal stages are given by

$$D_n = A_2 + ahA_1,$$

$$D_n k_{xi} = h[A_2(y_n + \sum_{j=1}^{i-1} \beta_{ij} k_{yj}) - F(x_n + \sum_{j=1}^{i-1} \beta_{ij} k_{xj}, y_n + \sum_{j=1}^{i-1} \beta_{ij} k_{yj})] +$$

$$+\eta A_2 \sum_{j \in J_i} \alpha_{ij} k_{xj} + (1 - \eta)hA_1 \sum_{j \in J_i} \gamma_{ij} k_{xj},$$

$$k_{yi} = \frac{1}{ah}[k_{xi} - h(y_n + \sum_{j=1}^{i-1} \beta_{ij} k_{yj}) - \eta \sum_{j \in J_i} \alpha_{ij} k_{xj} - (1 - \eta)h \sum_{j \in J_i} \gamma_{ij} k_{yj}],$$

if $i \in M_k$ and

$$D_n k_{xi} = \eta A_2(k_{x(i-1)} + \sum_{j \in J_i} \alpha_{ij} k_{xj}) + (\eta - 1)hA_1(k_{x(i-1)} + \sum_{j \in J_i} \gamma_{ij} k_{xj}),$$

$$k_{yi} = \frac{1}{ah}(k_{xi} - \eta(k_{x(i-1)} + \sum_{j \in J_i} \alpha_{ij} k_{xj}) - (1 - \eta)h(k_{y(i-1)} + \sum_{j \in J_i} \gamma_{ij} k_{yj}).$$

when $i \in M_m \backslash M_k$. Here, a, μ_i, β_{ij}, α_{ij} and γ_{ij} are parameters defining properties of stability and accuracy (6), h is the integration step, A_1 and A_2 are matrices approximating the derivatives $F_{ny} = \partial F(x_n, y_n)/\partial y$ and $F_{nx} = \partial F(x_n, y_n)/\partial x$. In what follows we use the notation $c_{ij} = \beta_{ij} + \gamma_{ij}$, where $\gamma_{ij} = 0$ if $j \notin J_i$ and $\gamma_{i,i-1} = 1$ if $j \in M_m \backslash M_k$. The matrix D_n is non-singular because $\det F_y \neq 0$. For the systems of index 1 or 2 the stages of the method are given by

$$(E - ahA_1)k_{xi} - ahA_2 k_{yi} = \delta_i hf(x_n + \sum_{j=1}^{i-1} \beta_{ij} k_{xj}, y_n + \sum_{j=1}^{i-1} \beta_{ij} k_{yj}) +$$

$$+\eta \sum_{j \in J_i} \alpha_{ij} k_{xj} + (1 - \eta)h \sum_{j \in J_i} \gamma_{ij}(A_1 k_{xj} + A_2 k_{yj}), \qquad (7)$$

$$-aB_1 k_{xi} - aB_2 k_{yi} = \delta_i g(x_n + \sum_{j=1}^{i-1} \beta_{ij} k_{xj}, y_n + \sum_{j=1}^{i-1} \beta_{ij} k_{yj}) +$$

$$+(1 - \eta)h \sum_{j \in J_i} \gamma_{ij}(B_1 k_{xj} + B_2 k_{yj}), \qquad (8)$$

where A_1, A_2, B_1, B_2 are matrices approximating the derivatives

$$f_{nx} = \frac{\partial f(x_n, y_n)}{\partial x}, \ f_{ny} = \frac{\partial f(x_n, y_n)}{\partial y}, \ g_{nx} = \frac{\partial g(x_n, y_n)}{\partial x}, \ g_{ny} = \frac{\partial g(x_n, y_n)}{\partial y},$$

and $\delta_i = 1$ if $i \in M_k$, $\delta_i = 0$ if $i \in M_m \backslash M_k$.

Reversibility of the matrix D_n is ensured for systems of index 1 by the reversibility of the matrix g_y, while for systems of index 2 – by the matrix $g_x f_y$.

The parameter η equals to 0 or 1. At $\eta = 0$, the schemes are preferable for computations, since they require less multiplications of a matrix by vector, and at $\eta = 1$ the schemes are more convenient in implementation .

The main feature of the schemes presented when compared to the conventional methods [11–14] is that in (m, k)–schemes the function F is evaluated k times at each step, and the number of stages is equal to m, $m \geq k$. The given schemes can be considered as a special form of ROW-methods, in which the set of definition of the scheme parameters is given more exactly. This simplifies the analysis of the order conditions, and the study of the problem how to use the time-lagged matrix D_n are carried out. The linear system of algebraic equations, arising in calculation of stages, is solved by the LU-decomposition of the matrix D_n. At every step once decomposition of the matrix D_n is evaluated, the function of the right side of a differential problem k times is calculated, backward in the Gauss method m times is executed. For given m and k the cost of one step is completely determined, and numbers $m_1, \ldots, m_k$ do only distribute this costs inside the step.

Two implementations of (6) for the systems of index 1 will be further considered:

a) the matrix D_n is reevaluated at each integration step;

b) the matrix D_n and the integration step h similar to [5] are not changed at several steps, thus $D_n = D_{n+\vartheta}$, $h_n = h_{n+\vartheta}$, $-Q \leq \vartheta \leq 0$ where Q is the maximum number of steps in the time-lagging of matrices derivatives.

3   Convergence and Order Conditions

The local error of the scheme (6) when solving (3) is defined as the difference between the exact and the numerical solution provided the initial values are choosen on the exact solution

$$\delta_x(t) = x_1 \quad x(t + h), \quad \delta_y(t) = y_1 - y(t + h).$$

We recall that order of consistency with respect to x is p and with respect to y is q, if

$$\delta_x(t) = O(h^{p+1}), \quad \delta_y(t) = O(h^{q+1}).$$

The condition for the parameters of a scheme ensuring the required order consistency can be obtained by equating the coefficients of the expansion of the approximate solution x_{n+1}, y_{n+1} to the exact solution

$$x(t_n + h) = x_n + \sum_{r=1}^{\infty} \frac{h^r}{r!} \sum_{\substack{t \in LT1X \\ \rho(t) = r}} [F(t)]_n, \tag{9}$$

$$y(t_n + h) = y_n + \sum_{r=1}^{\infty} \frac{h^r}{r!} \sum_{\substack{\mathbf{t} \in \mathbf{LT1Y} \\ \rho(\mathbf{t})=r}} [F(\mathbf{t})]_n, \qquad (10)$$

where $[F(\mathbf{t})]_n$ denotes a value of the elementary differential $F(\mathbf{t})$ of the order $\rho(\mathbf{t})$ at a point (x_n, y_n). Expressions (9), (10), trees set definition $\mathbf{T1} = \mathbf{T1X} \cup \mathbf{T1X}$, and the corresponding elementary differentials $F(\mathbf{t})$, $\mathbf{t} \in \mathbf{LT1}$ were introduced in [3].

Now we find an expansion similar to (9), (10) for the numerical solution at (x_{n+1}, y_{n+1}) for our scheme (6).

Assume that the (m, k)-scheme is implemented with time-lagging of the matrices derivatives. The following proposition gives the derivatives with respect to h at $h = 0$ of the entries of the matrix

$$\begin{bmatrix} f_x(x_n + \vartheta h) & f_y(y_n + \vartheta h) \\ g_x(x_n + \vartheta h) & g_y(y_n + \vartheta h) \end{bmatrix}$$

at a point (x_n, y_n).

Proposition 1. *Let $p \equiv f \vee g$ and $r \equiv x \vee y$. Then*

$$p_r^{(q)}(x_{n+\vartheta}, y_{n+\vartheta})|_{h=0} = \sum_{\substack{\mathbf{t} \in \mathbf{LT1X} \\ \rho(\mathbf{t})=q}} \vartheta^q [A_{pr}(\mathbf{t})]_n, \qquad (11)$$

where $[A_{pr}(\mathbf{t})]_n$ is a value of the differential

$$\frac{\partial^{k+l+1} p}{\partial r \partial x^k \partial y^l}(F(\mathbf{t}_1), \cdots, F(\mathbf{t}_k), F(\mathbf{u}_1), \cdots, F(\mathbf{u}_l)),$$

$\mathbf{t} = [\mathbf{t}_1, \cdots, \mathbf{t}_k, \mathbf{u}_1, \cdots, \mathbf{u}_l] \in \mathbf{LT1}$ *in the point (x_n, y_n).*

Differentiating p_r with respect to t gives

$$\frac{d^q p_r(x_n, y_n)}{dt^q} = \sum_{\substack{\mathbf{t} \in \mathbf{LT1X} \\ \rho(\mathbf{t})=q}} \frac{\partial^{k+l+1} p}{\partial r \partial x^k \partial y^l}(x^{(\alpha_1)}, \cdots, x^{(\alpha_k)}, y^{(\beta_1)}, \cdots, y^{(\beta_l)}).$$

Substituting of the expression for $x^{(\alpha_i)}, \cdots, y^{(\beta_j)}$ obtained from (9), (10) using the change of variables $dt = \vartheta dh$, gives the stated result as $h \to 0$.

We denote

$$\mathbf{t} = [\mathbf{t}_1^{\lambda_1}, \cdots, \mathbf{t}_n^{\lambda_n}]_r, \qquad (12)$$

for the tree $\mathbf{t} \in \mathbf{T1}$, where the index λ_i is the multiplicity of a inclusion of a corresponding subtree $\mathbf{t}_i \in \mathbf{T1}$, $r \equiv x \vee y$.

The number of a possible labelling $\alpha(\mathbf{t})$ of the tree $\mathbf{t} \in \mathbf{T1}$ is defined recursively by $\alpha(\mathbf{t}) = 1$, if

$$\rho(\mathbf{t}) = 1, \; \alpha(\mathbf{t}) = \bar{\rho}(\mathbf{t}) \prod_{i=1}^{n} \frac{1}{\lambda_i!} \left(\frac{\alpha(\mathbf{t}_i)}{\rho(\mathbf{t}_i!)} \right)^{\lambda_i},$$

where $\bar{\rho} = (\rho(\mathbf{t}) - 1)!$, if $\mathbf{t} \in \mathbf{T1X}$, $\bar{\rho} = \rho(\mathbf{t})!$, if $\mathbf{t} \in \mathbf{T1Y}$.

The integer number $\Gamma(\mathbf{t})$ corresponding to a tree $\mathbf{t} \in \mathbf{T1}$ is defined recursively by

$$\Gamma(\mathbf{t}) = 1, \; \text{if} \; \rho(\mathbf{t}) = 1,$$

$$\Gamma(\mathbf{t}) = \rho(\mathbf{t}) \prod_{i=1}^{n} \Gamma(\mathbf{t}_i)^{\lambda_i}, \; \text{if} \; \mathbf{t} = [\mathbf{t}_1^{\lambda_1}, \cdots, \mathbf{t}_n^{\lambda_n}]_x,$$

$$\Gamma(\mathbf{t}) = \prod_{i=1}^{n} \Gamma(\mathbf{t}_i)^{\lambda_i}, \; \text{if} \; \mathbf{t} = [\mathbf{t}_1^{\lambda_1}, \cdots, \mathbf{t}_n^{\lambda_n}]_y.$$

We put $\tilde{c}_{ij} = c_{ij}$, if $i > j$, $\tilde{c}_{ii} = a$, $\tilde{c}_{ij} = 0$, if $i < j \cdot \omega = (\omega_{ij})$ is the inverse of the matrix $(\tilde{c}_{ij})$.

The expression $\phi_i(\mathbf{t}) = \phi_{1i}(\mathbf{t}) + \phi_{2i}(\mathbf{t})/\Gamma(\mathbf{t})$, $\mathbf{t} \in \mathbf{T1}$, $1 \leq i \leq m$, is defined recursively by

$$\phi_{1i}(\mathbf{t}) = \delta_i, \; \phi_{2i}(\mathbf{t}) = 0,$$

if $\rho(\mathbf{t}) = 1$,

$$\phi_{1i}(\mathbf{t}) = \delta_i \prod_{r=1}^{n} \left(\sum_{\nu_r=1}^{i-1} \beta_{i\nu_r} \phi_{\nu_r}(\mathbf{t}_r) \right)^{\lambda_r},$$

$$\phi_{2i}(\mathbf{t}) = \rho(\mathbf{t}) \sum_{j=1}^{i} \gamma_{ij} \sum_{r=1}^{n} (\lambda_r \Gamma(\mathbf{t}_r) \vartheta^{(\rho(\mathbf{t}) - \rho(\mathbf{t}_r) - 1)} \phi_j(\mathbf{t}_r)),$$

if $\mathbf{t} = [\mathbf{t}_1^{\lambda_1}, \cdots, \mathbf{t}_n^{\lambda_n}]_x$,

$$\phi_{1i}(\mathbf{t}) = \sum_{j=1}^{i} \omega_{ij} \delta_j \prod_{r=1}^{n} \left(\sum_{\nu_r=1}^{j-1} \beta_{j\nu_r} \phi_{\nu_r}(\mathbf{t}_r) \right)^{\lambda_r},$$

$$\phi_{2i}(\mathbf{t}) = \sum_{1 \leq v \leq j \leq i} \omega_{ij} \gamma_{jv} \sum_{r=1}^{n} (\lambda_r \Gamma(\mathbf{t}_r) \vartheta^{(\rho(\mathbf{t}) - \rho(\mathbf{t}_r))} \phi_v(\mathbf{t}_r)),$$

if $\mathbf{t} = [\mathbf{t}_1^{\lambda_1}, \cdots, \mathbf{t}_n^{\lambda_n}]_y$.

The expansion of the derivatives of the numerical solution is given by the following proposition.

Proposition 2.

$$k_{xi}^{(q)} = \sum_{\substack{\mathbf{t} \in \mathbf{LT1X} \\ \rho(\mathbf{t}) = q}} \Gamma(\mathbf{t}) \phi_i(\mathbf{t}) [F(\mathbf{t})]_n, \tag{13}$$

$$k_{yi}^{(q)} = \sum_{\substack{t \in \mathbf{LT1Y} \\ \rho(t)=q}} \Gamma(t)\phi_i(t)[F(t)]_n. \tag{14}$$

This proposition generalizes the theorem (4.4) from [3] and , for $q = 1$, (13), (14) coincide with the corresponding expressions from [3].

The order conditions are defined by the following proposition.

Proposition 3.

$\delta_n^x = O(h^{p+1})$, *if the conditions* $\sum\limits_{i=1}^{m} \mu_i \phi_i(t) = \frac{1}{\Gamma(t)}$,

hold for all trees $t \in \mathbf{T1X}$ *of order* $\rho(t) \leq p$,

$\delta_n^y = O(h^{q+1})$, *if* $\sum\limits_{i=1}^{m} \mu_i \phi_i(t) = \frac{1}{\Gamma(t)}$ *hold*,

for all trees $t \in \mathbf{T1Y}$ *of order* $\rho(t) \leq q$.

A numerical solution converges with order p with respect to x and with order q with respect to y if the global error

$$e_n^x = x_n - x(t_n), \quad e_n^y = y_n - y(t_n)$$

satisfies

$$e_n^x = O(h^p), \quad e_n^y = O(h^q).$$

Applying methods (6) for solving the scalar test equation $x' = \lambda x$ we obtain $x_{n+1} = R(z)x_n$, $z = h\lambda$, where $R(z)$ is called a stability function.

The following theorem answers the question on convergence of the (m, k)-methods (6).

Proposition 4. *Assume that scheme (6) is consistent of order p with respect to x and of order $(q - 1)$ with respect to y Suppose that the stability factor is such that $|R(\infty)| < 1$ (stability function at ∞). Then numerical solution converges to the exact solution with the order $\bar{p}$ on variable x and with the order q on variable y, where the value $\bar{p}$ is set by above chosen implementation of the scheme a) or b):*

$$a) \ \bar{p} = \min(p, 2q), \ b) \ \bar{p} = \min(p, q + 1).$$

We note, that the given proposition in the case p=q follows from the theorem 1 [3] true for a wider class of the one-step methods.

In Tables 1, 2 the order conditions ensuring convergence of (m, k)-methods up to the fourth order of accuracy are tabulated. We use the notations

$$\gamma_i = \sum \gamma_{ij}\delta_j, \ \tilde{c}_i = \sum \tilde{c}_{ij}\delta_j, \ \beta_i = \sum \beta_{ij}\delta_j.$$

Table 1. Order conditions for the x-component

$\rho(t)$	t		
1	•	$\sum \mu_i \delta_i = 1$	(15.a)
2		$\sum \mu_i \tilde{c}_i = \frac{1}{2}$	(15.b)
3		$\sum \mu_i \beta_i^2 + 2\vartheta \sum \mu_i \gamma_i = \frac{1}{3}$	(15.c)
3		$\sum \mu_i \tilde{c}_{ij} \tilde{c}_j = \frac{1}{6}$	(15.d)
4		$\sum \mu_i \beta_i^3 + 3\vartheta \sum \mu_i \gamma_i = \frac{1}{4}$	(15.e)
4		$\sum \mu_i \beta_i \beta_{ij} \tilde{c}_j + \vartheta \sum \mu_i \gamma_{ij} \tilde{c}_j + \frac{1}{2}\vartheta^2 \sum \mu_i \gamma_i = \frac{1}{8}$	(15.f)
4		$\sum \mu_i \tilde{c}_{ij} \beta_j^2 + 2\vartheta \sum \mu_i \tilde{c}_{ij} \gamma_j = \frac{1}{12}$	(15.g)
4		$\sum \mu_i \tilde{c}_{ij} \tilde{c}_{jk} \tilde{c}_k = \frac{1}{24}$	(15.h)
4		$\sum \mu_i \beta_i \beta_{ij} \omega_{jk} \beta_k^2 + \vartheta \sum \mu_i (2\beta_i \beta_{ij} \omega_{jk} \gamma_k + \gamma_{ij} \omega_{jk} \beta_k^2) + \vartheta^2 \sum \mu_i (\gamma_i + 2\gamma_{ij} \omega_{jk} \gamma_k) = \frac{1}{4}$	(15.i)

Table 2. Order conditions for the y - component

$\rho(t)$	t		
2		$\sum \mu_i \omega_{ij} \beta_j^2 + 2\vartheta \sum \mu_i \omega_{ij} \gamma_j = 1$	(15.j)
3		$\sum \mu_i \omega_{ij} \beta_j^2 + 3\vartheta^2 \sum \mu_i \omega_{ij} \gamma_j = 1$	(15.k)
3		$\sum \mu_i \omega_{ij} \beta_j \beta_{jk} \tilde{c}_k + \vartheta \sum \mu_i \omega_{ij} \gamma_{jk} \tilde{c}_k + \frac{1}{2}\vartheta^2 \sum \mu_i \omega_{ij} \gamma_j = \frac{1}{2}$	(20.l)
3		$\sum \mu_i \omega_{ij} \beta_j \beta_{jk} \omega_{ks} \beta_s^2 + \vartheta \sum \mu_i \omega_{ij} (2\beta_j \beta_{jk} \omega_{ks} \gamma_s + \gamma_{jk} \omega_{ks} \beta_s^2) + \vartheta^2 \sum \mu_i \omega_{ij} (\gamma_j + 2\gamma_{jk} \omega_{ks} \gamma_s) = 1$	(15.m)

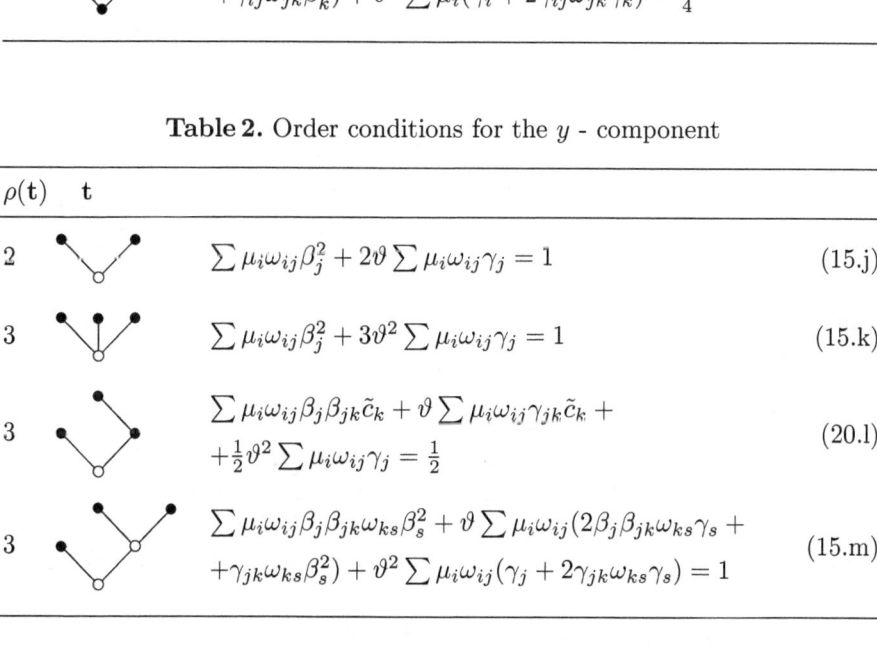

4   (m,k)-Schemes of the Optimum Order

We study the utmost achievable order of accuracy by (m, k)-schemes for given $k \leq 4$ for system (1) of index 1. First we consider the case of the implementation a) of the scheme (6).

Let $k = 1$ and let us consider the schemes with one evaluation of the function F at a step. In the case $m = 1$ the stability function takes the form $R(z) = [1 + (\mu_1 - a)z]/(1 - az)$. Under $\mu_1 = 1$, $a = 0.5$ the order conditions of the second order are satisfied. However, unlike ODE systems, the scheme has only the first order of accuracy, as far as $|R(\infty)| = 1$. Under $\mu_1 = a = 1$ we have the L-stable $(R(\infty) = 0)$ scheme of the first order, which in [6] is applied to solve the problem of index 0.

In the case $m = 2$ the conditions of the second order yield $\mu_1 = 1$, $\mu_2 = 0.5a$, and

$$R(z) = \frac{1 + (1 - 2a)z + (0.5 - 2a + a^2)z^2}{(1 - az)^2}, \quad R(\infty) = \frac{0.5 - 2a - a^2}{a^2}.$$

Setting $a = 1 - 0.5\sqrt{2}$ or $a = 1 + 0.5\sqrt{2}$ we obtain the parameters of L-stable $(2, 1)$-scheme of the second order.

Proposition 5. *For all m there exists no $(m, 1)$-method of order higher than 2.*

The given proposition is a consequence an analogous statement from [9].

Let $k = 2$ and we consider the schemes with two evaluation of the function F on a step. Easily to be convinced, that at $m = 2$ the maximum order is equal to 2. In the case $m = 3$, $M_2 = \{1, 2\}$ the conditions of the third order imply

$$\mu_1 = \beta_{21}^{-2}(3\beta_{21}^2 - 1)/3, \quad \mu_2 = \beta_{21}^{-2}/3, \quad \mu_3 = \beta_{21}^{-2}(a - 3a^2)/3,$$

$$c_{21} = \frac{-6a^2 + 6a - 1}{6a^2 - 2a}\beta_{21}^2, \quad c_{31} = \frac{18a^3 - 21a^2 + 9a - 1}{18a^4 - 12a^3 + 2a^2}\beta_{21}^2 - 1,$$

where a and β_{21} are free parameters. Under $1/3 \leq a \leq 1.068579$ [12] a scheme is A-stable, and under $a \approx 0.43587$ (i.e. a is root of the $a^3 - 3a^2 + 2a/3 - 1/6 = 0$) a scheme is L-stable.

Proposition 6. *For all m and for any choice of sets (5) there exists no $(m, 3)$-method of order higher than 3 for the y-component.*

Let $k = 3$, $M_3 = \{1, s, r\}$, $1 < s < r \leq m$. We denote

$$q_s = \sum_{i=s}^{m} \mu_i \omega_{ij}, \quad q_r = \sum_{i=r}^{m} \mu_i \omega_{ij}, \quad u_r = \sum_{r>j\geq l}^{m} \beta_{rj}\omega_{jl}\beta_l^2.$$

The conditions of the fourth order (15.c), (15.e), (15.j), (15.k), (15.i), (15.m) yields

$$\mu_s \beta_s^2 + \mu_r \beta_r^2 = \frac{1}{3}, \quad \mu_s \beta_s^3 + \mu_r \beta_r^3 = \frac{1}{4}, \quad q_s \beta_s^2 + q_r \beta_r^2 = 1,$$

$$q_s \beta_s^3 + q_r \beta_r^3 = 1, \quad \mu_r \beta_r u_r = \frac{1}{4}, \quad q_r \beta_r u_r = 1.$$

We introduce the matrices

$$A = \left\{ \begin{matrix} \mu_s \ \mu_r \\ q_s \ q_r, \end{matrix} \right\}, \quad B = \left\{ \begin{matrix} \beta_s^2 \ \beta_s^3 \\ \beta_r^2 \ \beta_r^3 \end{matrix} \right\}, \quad C = \left\{ \begin{matrix} 1/3 \ 1/4 \\ 1 \ \ 1 \end{matrix} \right\}, \quad D = \left\{ \begin{matrix} 1 \ 0 \\ -4 \ 1 \end{matrix} \right\},$$

then the first four equations can be represented in the form of the matrix equality: $AB = C$. We notice that $\beta_s \neq 0$, as far as $\det(C) \neq 0$. The last two equations give $q_r = 4\mu_r$. Multiplying the matrix equality from the right-hand-side by matrix B^{-1} and from the left by D, we have for the right bottom element of the product

$$0 = \beta_s/(3\beta_r^2(\beta_s - \beta_r)).$$

The obtained contradiction proves the proposition.

However for the explicit problem (2) it is possible to obtain the methods of the fourth order, in addition ensuring the third order for the problem (4) of index 1. In the case $m = 4$, $M_2 = \{1, 3\}$ the parameters of the A-stable scheme are

$$a = \frac{1}{2}, \ \mu_1 = \frac{11}{27}, \ \mu_2 = -\frac{8}{27}, \ \mu_3 = \frac{16}{27}, \ \mu_4 = -\frac{4}{27},$$

$$\beta_{31} = \frac{3}{4}, \ \beta_{32} = -\frac{3}{32}, \ c_{32} = -\frac{9}{32}, \ c_{42} = -\frac{21}{16}.$$

and parameters of the L-stable scheme at $m = 5$, $M_2 = \{1, 3\}$ are

$$\mu_1 = \frac{11}{27}, \ \mu_2 = \frac{-22a + 5}{54}, \ \mu_3 = \frac{16}{27}, \ \mu_4 = \frac{-16a + 4}{27},$$

$$\mu_5 = \frac{48a^3 - 32a^2 + 4a}{27}, \ \beta_{31} = \frac{3}{4}, \ \beta_{32} = \frac{-24a + 9}{32},$$

$$c_{32} = \frac{216a^4 - 864a^3 + 648a^2 - 144a + 9}{384a^2 - 256a + 32},$$

$$c_{52} = \frac{-6912a^6 + 16416a^5 - 14832a^4 + 6296a^3 - 1263a^2 + 114a - 4}{6912a^6 - 13824a^5 + 10944a^4 - 4352a^3 + 912a^2 - 96a + 4},$$

$$c_{42} = \left[c_{52}(576a^5 - 768a^4 + 352a^3 - 64a^2 + 4a) - \right.$$

$$\left. -216a^4 + 4a^3 + 159a^2 - 45a - 3 \right]/(192a^3 - 176a^2 + 48a - 4),$$

where a is choosen such that $0.2479 < a < 0.67604$ [12].

Note, that the properties of stability of (m, k)-methods depend on the choice of the set M_k. The following proposition in particular holds.

Proposition 7. *There exists a L-stable $(4, 3)$-scheme of order 4 with respect to x and of order 3 with respect to y.*

However, the study of methods at $M_3 = \{1, 2, 3\}$ shows that $|R(z)| > 1$. If we consider the case $M_3 = \{1, 2, 4\}$, the parameters of the L-stable scheme are the following:

$$\mu_4 = \frac{4\beta_2 - 3}{12\beta_4^2(\beta_2 - \beta_4)}, \ \mu_2 = \frac{1 - 3\mu_4\beta_4}{3\beta_2}, \ \mu_1 = 1 - \mu_2 - \mu_4,$$

$$c_{21} = \frac{(-24a^3 + 36a^2 - 12a + 1)\beta_2}{24a^3 - 16a^2 + 2a}, \ c_{43} = \frac{12a^3 - 8a^2 + a}{12\mu_4\beta_2^2},$$

$$c_{31} = \frac{(-12\mu_4 c_{43} + 12a^2 - 12a + 2)\beta_2^2 + (4a - 1)c_{21}}{12\mu_4 c_{43}\beta_2^2},$$

$$\beta_{43} = \frac{(-8a^3 + 3a)\beta_2 - 6a^2 c_{21}}{24\mu_4(c_{21} + ac_{31} + a)},$$

$$\beta_{42} = \frac{4\mu_4\beta_4\beta_{43} + a}{4a\mu_4\beta_2\beta_4}, \ \beta_{41} = \beta_4 - \beta_{42},$$

$$\mu_3 = \frac{-12\mu_4\beta_{42}\beta_2^2 - 4a + 1}{12\beta_2^2},$$

where β_2 and β_4 are free parameters, $a \approx 0.572816$.

Proposition 8. *There exist embedded $(5, 4)$-schemes of order 4 and 3 determined by the set $M_4 = \{1, 3, 4, 5\}$. The scheme of order 4 is L-stable and the scheme of order 3 is A-stable.*

Let β_3, β_4, β_5, β_{32}, β_{54}, c_{54} be, in general, free parameters. We use a short notation

$$q_s = \sum_{i=s+1}^{5} \mu_i\omega_{ij}, \ u_s = \sum_{s>j\geq l}^{m} \beta_{sj}\omega_{jl}\beta_l^2, \ s = 3, 4, 5.$$

The conditions of the fourth order (15.j), (15.k), (15.i), (15.m) yields

$$q_3\beta_3^2 + q_4\beta_4^2 = 1 - \frac{1}{3a},$$

$$q_3\beta_3^3 + q_4\beta_4^3 = 1 - \frac{1}{4a},$$

$$\mu_4\beta_4 u_4 + \mu_5\beta_5 u_5 = \frac{1}{4}, \ q_4\beta_4 u_4 = 1.$$

Having chosen the free parameters, we obtain q_3, q_4 from the first two equations and μ_5 from the expression $q_4 = -a^{-2}c_{54}\mu_5$.

Equations (15.c), (15.e)

$$\mu_3\beta_3^2 + \mu_4\beta_4^2 + \mu_5\beta_5^2 = \frac{1}{3},$$

$$\mu_3\beta_3^3 + \mu_4\beta_4^3 + \mu_5\beta_5^3 = \frac{1}{4}$$

give μ_3, μ_4.

Now u_4, u_5 are obtained from (15.i), (15.m). Using $u_4 = \beta_3^2\beta_{43}$ we get β_{43}. Parameters c_{43}, c_{53} are obtained from the expression

$$q_3 = a^{-3}\big[\mu_5(c_{43}c_{54} - ac_{53}) - a\mu_4 c_{43}\big]$$

and from the equation (15.g)

$$\mu_4\beta_3^2 c_{43} + \mu_5\big(\beta_3^2 c_{53} + \beta_4^2 c_{54}\big) = \frac{1}{12} - \frac{a}{3}.$$

The expression

$$u_5 = a^{-2}\big[(a\beta_{53} - c_{43}\beta_{54})\beta_3^2 + a\beta_{54}\beta_4^2\big]$$

gives β_{53}.

From the conditions (15.l), (15.f)

$$q_3\beta_3\beta_{32} + q_4\beta_4(\beta_{43}\beta_3 + \beta_{42}) = \frac{5}{6} - \frac{1}{8a},$$

$$\mu_3\beta_3\beta_{32} + \mu_4\beta_4(\beta_{42} + \beta_{31}\beta_{43}) + \mu_5\beta_5(\beta_{52}+$$

$$+\beta_{54}c_{43} + \beta_{41}\beta_{54} + \beta_{31}\beta_{53}) = \frac{1}{8} - \frac{a}{3},$$

the equation

$$\mu_5 c_{32}c_{43}c_{54} = a^5 - 4a^4 + 3a^3 - \frac{2a^2}{3} + \frac{a}{24},$$

ensuring the L-stability, and from the conditions (15.h), (15.d), (15.b), (15.a)

$$\mu_4 c_{32}c_{43} + \mu_5(\beta_3 c_{43}c_{54} + c_{32}c_{53} + c_{42}c_{54}) = \frac{1}{24} - \frac{a}{2} + \frac{3a^2}{2} - a^3,$$

$$\mu_3 c_{32} + \mu_4(\beta_3 c_{43} + c_{42}) + \mu_5(\beta_3 c_{53} + \beta_{41}c_{54} + c_{43}c_{54} + c_{52}) = \frac{1}{6} - a + a^2,$$

$$\mu_2 + \mu_3\beta_{31} + \mu_4(\beta_{41} + c_{43}) + \mu_5(\beta_{51} + c_{53} + c_{54}) = \frac{1}{2} - a,$$

$$\mu_1 + \mu_3 + \mu_4 + \mu_5 = 1$$

we evaluate sequentially the parameters β_{42}, β_{52}, c_{32}, c_{42}, c_{52}, μ_1, μ_2.

Degeneration of the minor

$$\left\{ \begin{matrix} \beta_3^2 & \beta_4^2 & \beta_5^2 \\ c_{32} & \beta_3 c_{43} + c_{42} & c_{52} + \beta_3 c_{53} + c_{54}(\beta_{41} + c_{43}) \\ 0 & -a\beta_3^2 c_{43} & -a\beta_3^2 c_{53} + c_{54}(\beta_3^2 c_{43} - a\beta_4^2) \end{matrix} \right\},$$

corresponding to the parameters $\tilde{\mu}_3$, $\tilde{\mu}_4$, $\tilde{\mu}_5$ in the order conditions (15.c), (15.d), (15.j) of the embedded scheme, ensures the existence of embedded method of the order 3. This condition gives the algebraic equation at parameter a

$$84a^4 - 132a^3 + 72a^2 - 15a + 1 = 0$$

having two of the solutions in $\mathbf{R}$: $a_1 \approx 0.130354$, $a_2 \approx 0.239192$.

Choosing the second value a and setting the free parameter $\tilde{\mu}_5$ we obtain from conditions (15.a) – (15.d) of the embedded scheme other values of the parameters $\tilde{\mu}_i$.

Now we consider the implementation b) with the time-lagging of matrices derivatives. Assume that the coefficients of the scheme are independent of the parameter ϑ. In the case of order 3 accuracy this yields the two additional order conditions

$$\sum \mu_i \beta_i = \frac{1}{2}, \tag{15.c'}$$

$$\sum \mu_i \omega_{ij} \beta_j = 1. \tag{15.j'}$$

Proposition 9. *For all m there exists no $(m, 2)$-method of order 3 satisfying* (15.c'), (15.j').

This follows from the inconsistency of (15.c'), (15.j'), and (15.j).

In the case $m = 3$, $k = 3$ the parameters of the L-stable scheme are

$$\mu_1 = \frac{(6a - 1)\beta_3 - 2a}{4(3a - 1)\beta_3}, \quad \mu_3 = -\frac{a}{\beta_3((6a - 3)\beta_3 - 6a + 2)},$$

$$\mu_2 = \frac{(6a - 3)(1 - 2\mu_3\beta_3)}{4(3a - 1)}, \quad \beta_2 = \frac{6a - 2}{6a - 3}, \quad \beta_{32} = \frac{a(1 - 2a)}{2\mu_3\beta_2},$$

$$c_{21} = \frac{6a^2 - 6a + 1}{6\mu_3 c_{32}}, \quad c_{31} = \frac{1 - 2\mu_2 c_{21} - 2\mu_3 c_{32} - 2a}{2\mu_3},$$

where $a \approx 0.43587$, and β_3, β_{32} are free parameters.

In addition in the case of the order 4 accuracy it is necessary to satisfy 7 conditions

$$\sum \mu_i \beta_{ij} c_j = \frac{1}{6} - \frac{a}{2}, \tag{15.f'}$$

$$\sum \mu_i c_{ij} \beta_j = \frac{1}{6} - \frac{a}{2}, \tag{15.g'}$$

$$2\sum \mu_i \beta_i \beta_{ij} \omega_{jk} \beta_k + \sum \mu_i \beta_{ij} \omega_{jk} \beta_k^2 = 1, \tag{15.i'}$$

$$\sum \mu_i \beta_{ij} \omega_{jk} \beta_k = \frac{1}{2}, \tag{15.i''}$$

$$\sum \mu_i \tilde{\omega}_{ij} \beta_{jk} c_k = \frac{1}{2} - a - \frac{1}{6a}, \tag{15.l'}$$

$$2\sum \mu_i \tilde{\omega}_{ij} \beta_j \beta_{jk} \omega_{kl} \beta_l + \sum \mu_i \tilde{\omega}_{ij} \beta_{jk} \omega_{kl} \beta_l^2 = 3 - \frac{1}{a}, \tag{15.m'}$$

$$\sum \mu_i \tilde{\omega}_{ij} \beta_{jk} \omega_{kl} \beta_l = 1 - \frac{1}{2a}. \tag{15.m''}$$

Here is $\tilde{\omega} = \omega - a^{-1}I$, where I is the identity matrix.

The following result for the scheme with the time-lagging of matrices derivatives similar to Proposition 7 holds.

Proposition 10. *There exists an L-stable* $(10, 4)$*-scheme of order 4 accuracy in both variables with the time-lagging of matrices derivatives.*

We present this result without the proof, since the proof is too complicated.

In conclusion we note that at $m \leq 9$ there exists no the $(m, 4)$-scheme of the order 4 accuracy in both variables with the time-lagging of matrices derivatives.

References

1. Boyarincev, Y.A., Danilov, V.A., Loginov, A.A., Chistyakov, V.F.: The Numerical Methods for Singular Systems. Nauka, Novosibirsk (1989)
2. Gear, C.W.: Differential-algebraic equations index transformations. SIAM J. Sci. Stat. Comput. **9**, 39–47 (1988)
3. Roche, M.: Rosenbrock methods for differential algebraic equation. Numer. Math. **52**, 45–63 (1988)
4. Levykin, A.I., Novikov, E.A.: A study of (m, k)-methods of the order 3 for implicit systems of ordinary differential equations. Novosibirsk, Preprint Computing Center SB RAS **882** (1990)
5. Werwer, J.G., Scholz, S., Blom, J.G., Louter-Nool, M.: A class Runge-Kutta-Rosenbrock Methods for solving stiff differential equations. ZAAM **63**, 13–20 (1983)
6. Novikov, E.A., Yumatova, L.A.: Some onestep methods for solving differential algebraic equation. Novosibirsk, Preprint Computing Center SB RAS **661** (1986)
7. Dekker, K., Verver, J.G.: Stability of Runge-Kutta Methods for Stiff Nonlinear Differential Equations. North-Holland, Amsterdam (1984)
8. Levykin, A.I., Novikov, E.A.: (m, k)-method of the order 2 for implicit systems of ordinary differential equations. Novosibirsk, Preprint Computing Center SB RAS **768** (1987)
9. Novikov, E.A., Shitov, Y.A., Shokin, Y.I.: A class of (m, k)-methods for solving stiff systems. Dokl. Akad. Nauk. **301**(6), 1310–1313 (1988)
10. Elsgolts, L.E.: Differential Equation and Variatsional Calculus. Nauka, Moskow (1969)
11. Deuflhard, P., Hairer, E., Zugck, J.: One-step and extrapolation methods for differential-algebraic systems. Numer. Math. **51**, 501–516 (1987)
12. Hairer, E., Wanner, G.: Solving Ordinary Differential Equations II: Stiff and Differential-Algebraic Problems. Springer, Heidelberg (1991)
13. Mao, G., Petzold, L.R.: Efficient integration over discontinuities for differential-algebraic systems. Computers Mathematics with Applications **43**(1), 65–79 (2002)
14. Boscarino, S.: Error Analysis of IMEX RungeKutta Methods Derived from Differential-Algebraic Systems. SIAM J. Numer. Anal. **45**(4), 1600–1621 (2007)

Application of Immersed Boundary Method in Modelling of Thrombosis in the Blood Flow

Saule Maussumbekova[✉] and Assel Beketaeva

Al-Farabi Kazakh National University, av. al-Farabi, 71, 050008 Almaty, Kazakhstan
saulemaussumbekova@gmail.com, azimaras@mail.ru
http://www.kaznu.kz

Abstract. Thrombosis occurrence is associated with hemodynamics instability. For prediction of it various experimental and numerical methods are developed. However, the greatest interest is mathematical methods for computing the hemodynamic parameters in thrombus formation. The model is possible to calculate the basic hemodynamic parameters of blood flow and the development of stenosis as a result of thrombosis. To describe the two dimensional blood flow in vessels with complex geometry as incompressible Newtonian fluid was used the conservation momentum law. Changing the shape of the vascular bed is considered in connection with possible biochemical processes like blood clots. It was assumed that convective flows do not have significant changes with the growth of blood clots, however, it is not conclusive with respect to real systems. Thrombus growth entails a change in the flow region, which is taken into account in this study using the immersed boundary method. The presence of the immersed boundary is taken into account by adding a special function in the equation of motion, allowing you to accurately represent streamlined border area. Unknown special function determined at the numerical solution stage of the problem, thus removing the requirement elastic boundaries. Also model consists from the equations describing the dynamics of the distribution of the main metabolites of blood clotting. For the numerical solution of the problem the method of splitting into physical parameters was used. To approximate the convective terms were used the quasi monotone high-order schemes. As a result of numerical experiments it was found that the use of the immersed boundary method qualitatively describes the dynamics of the stenosis as a result of thrombosis.

Keywords: Hemodynamics · Incompressible newtonian fluid · Navier-stokes equations · Complex geometry · The immersed boundary method · Numerical solution · Thrombus

1 Introduction

Modern medicine is essentially an experimental science with great empirical impact experience on the course of disease by various means. Experimental study

© Springer International Publishing Switzerland 2015
N. Danaev et al. (Eds.): CITech 2015, CCIS 549, pp. 108–116, 2015.
DOI: 10.1007/978-3-319-25058-8_11

is limited for a detailed study of the processes in biological media and the mathematical modeling is the most effective method for their research. Statements of the biological and medical problems that lead to the need for the numerical solution of systems of partial differential equations appeared relatively recently. They are presented in [1]. For the numerical solution of these problems were used methods previously used for solving fluid physics [2]. The rheological relations for the biological mechanics are developed in [3]. Description of the simplest mathematical model of the circulatory of heart can be found in [4]. The functions of the circulatory system of humans, which consists of small and large circles, are very important and diverse, so their modeling, both in normal and pathological conditions, is one of the biggest challenges of medicine. The dynamic model of pulsating fluid flows in the expandable tubes is the most adequate. Quasi-one dimensional - hydraulic model of an incompressible fluid in a deformable blood vessel, in the case of a generalized to hierarchical branching of factorial structured blood vessels were used in [5]. Another approach of modeling the functioning of the circulatory system based on a quasi three-dimensional circulatory system was proposed in [6]. In this case, The change of all parameters which can be output is a subject to simulation, for example, the concentration of active substances in the blood and the pressure on different parts of the circulatory system, as well as the velocity of blood flow. Normal functioning of the blood coagulation system provides liquid flow state of blood. The ability to provide a rapid local reaction of the body in response to the local disruption of normal flow conditions is a feature of the blood coagulation system. Maintaining the integrity of the circulatory system is provided by high speed of activation reactions of the coagulation system. Cascading biochemical mechanisms of signal amplification from of the lesion of the vascular wall provide speed of these reactions [7]. To date, the number of mathematical models describing the kinematics of the activation of key metabolites of the blood coagulation system were developed [8–11]. Interaction processes of blood coagulation with mass transfer are given in [25]. Effect of convective transport in the distribution of the factors of the coagulation system in space, affecting the growth of the thrombus is analyzed in [26], but the effect of a growing thrombus on flow around it was not taken into account. In this paper we formulate a mathematical model describing the production of the main metabolites of the coagulation system, their transport and distribution of the flow by diffusion. The qualitative dependences of activation thrombotic conditions in blood flow were investigated from its properties such as viscosity, flow conditions as pressure drop and chemical composition.

2   Mathematical Model

The formation of a dense fibrin polymer which prevents substance migration is a blood thrombus in the vessel. The flow is not only involved in the transfer of key metabolites of blood clotting in the space, but also has a direct effect on blood clot, deforming it. The transition from one type of spatial temporal behavior to a qualitatively different type is of great interest in the study of the dynamics of growth of blood clots in the bloodstream, it means to define the conditions under which the activation coagulation system is associated by damage involve the formation of a blood clot, and in which they are not sufficient to stimulate thrombus formation. The flow of blood is described by the equations of a viscous incompressible fluid. As a place of activation of a cascade of biochemical reactions will be the portion of the surface of the local damage to the vessel wall. The size of the damage and the intensity of activation of coagulation in this area are free parameters of the problem. Viscous flow described by non-stationary Navier-Stokes equations.

$$div\,\boldsymbol{V} = 0 \tag{1}$$

$$\frac{\partial \boldsymbol{V}}{\partial t} + (\boldsymbol{V}\nabla)\boldsymbol{V} = -\frac{1}{\rho}\nabla p + \nu \Delta \boldsymbol{V}, \tag{2}$$

$$\frac{\partial C_k}{\partial t} + \nabla(\boldsymbol{V_k}C_k - D_k\nabla C_k) = F_k(C_1, ..., C_m) \tag{3}$$

D_k- the diffusion coefficient of the k metabolite, C_k-- their concentration, $F_k(C_1, ..., C_m)$ the term describing the kinetics of local production of the substance, V_k-- its rate of convective transfer. It is assumed that the transfer rate of each of the main metabolites is given by $V_k = b_k V$, b_k, D_k are allowed to be zero for fibrin, and b_k is one for all soluble metabolites. The flow is missing where the polymer clot have a density above a certain value ψ^c. The velocity of the flow at the interface of the polymer clot is considered to be zero. We assume that in the area of injury inputted the activator, that is local increase in the activator concentration is given in the initial time. The equations describing production and decay of the spatial transfer of activator and inhibitor, as well as the production of fibrin are as follows [14].

$$\frac{\partial \theta}{\partial t} = D_1\Delta\theta - div\,(\boldsymbol{V}\theta) + \frac{\alpha\theta^2}{\theta + \theta_0} - \gamma\theta\phi - \chi_1\theta \tag{4}$$

$$\frac{\partial \phi}{\partial t} = D_2\Delta\phi - div\,(\boldsymbol{V}\varphi) + \beta\theta\left(1 + \frac{\phi^2}{\phi_0^2}\right) - \chi_2\phi \tag{5}$$

$$\frac{\partial \psi}{\partial t} = k\theta \tag{6}$$

The model describes the formation of the main regulators of fibrin polymerization in the blood - activator of clotting (thrombin), the concentration of which is noted by θ, the inhibitor concentration is indicated by ϕ. Thrombin catalyzes a reaction to convert the precursor of fibrin - fibrinogen into fibrin monomer, concentration of which is denoted by ψ, it in turn polymerizes in case and gives thrombus.

We could rewrite the equations (5), (6) cause of $div\,\boldsymbol{V} = 0$.

$$\frac{\partial \theta}{\partial t} = D_1\Delta\theta - \boldsymbol{V}\nabla\theta + \frac{\alpha\theta^2}{\theta + \theta_0} - \gamma\theta\phi - \chi_1\theta \tag{7}$$

$$\frac{\partial \phi}{\partial t} = D_2 \Delta \phi - \boldsymbol{V} \nabla \phi + \beta \theta \left(1 + \frac{\phi^2}{\phi_0^2} \right) - \chi_2 \phi \tag{8}$$

Thus, the mathematical model (2), (6), (7), (8) describes the change in the velocity field in the formation of thrombus in the vessel. To simulate the obstacles of arbitrary shape (in this problem blood clot) is introduced by a discrete-time artificial power. This force is applied only on the surface and within the constraints of the body. Force application point disposed in a spaced, similar velocity components defined on a staggered grid. When the point of application of force coincides with a virtual border, an artificial force is applied so as to satisfy the boundary conditions on the obstacle. The cell containing the virtual boundary, does not satisfy the equation of conservation of mass. Therefore, we introduce the source / drain weight to the cell that contains the virtual border. Discrete in time force is used to meet the conditions of adhesion on a virtual border, while the source / drain weight, to meet the conservation of mass for the cell that contains the virtual boundary. Procedure nondimensionalization this system involves choosing the characteristic scales: the concentrations θ_0 and ϕ_0, lines size L , the characteristic scale of velocity $\boldsymbol{V}$. In view of the above equations (1) - (2) takes the form:

$$div\boldsymbol{V} - q = 0 \tag{9}$$

$$\frac{\partial \boldsymbol{V}}{\partial t} + (\boldsymbol{V}\nabla)\boldsymbol{V} = -\nabla p + \frac{1}{Re}\Delta \boldsymbol{V} \quad + f_i \tag{10}$$

where $Re = LV/\nu$- Reynolds number, f_i - components of artificial power defined on a cell boundary in a virtual boundary or within of the body ($f_i = (f_u, f_v)$), q- source / drain of weight defined in the center of the cell on the virtual border or inside the body. So

$$f_u = \left\{ \begin{matrix} 0, \ (x, z) \in \Omega/\Omega_0 \\ f_u, (x, z) \in \Omega_0 \end{matrix} \right\}, \quad f_v = \left\{ \begin{matrix} 0, (x, z) \in \Omega/\Omega_0 \\ f_v, (x, z) \in \Omega_0 \end{matrix} \right\}, \quad q = \left\{ \begin{matrix} 0, (x, z) \in \Omega/\Omega_0 \\ q, (x, z) \in \Omega_0 \end{matrix} \right\}$$

where Ω_0 - region of thrombus, Ω- region without thrombus. Note that the model due to the rigidity of the reaction part of the system is difficult for numerical implementation. The flows of matter were basic calculated values in the numerical implementation [15], which allows to build a conservative difference schemes for stiff systems. The characteristic scales of concentrations were used to nondimensionalization of model equations. Thus, the transfer equation of reagents will take the form:

$$\frac{\partial \theta}{\partial t} = \frac{1}{Pe}\Delta \theta - \boldsymbol{V}\nabla \theta + \frac{1}{M}\left(\frac{\theta \left(\theta - \overline{\chi_1} \right)}{\theta + 1} - \overline{\gamma}\theta\phi \right) \tag{11}$$

$$\frac{\partial \phi}{\partial t} = \frac{1}{Pe}\Delta \theta - \boldsymbol{V}\nabla \phi + \frac{1}{M}\left(b\theta \left(1 - \varepsilon\phi \right)\left(1 + \phi^2 \right) - \chi_2\phi \right) \tag{12}$$

where $M = \frac{V}{a_* L}$, $Pe = \frac{LV}{D}$, $\chi_1 = a_* \overline{\chi_1}$, $\chi_2 = a_* \overline{\chi_2}$, $b = \frac{\beta \theta_0}{\varphi_0 a_*}$, $c = \frac{\varphi_0}{\varepsilon}$. The value of the constants listed in [16].

The boundary conditions for the Navier-Stokes equations were taken as follows: on the walls of the vessel and the surface of a blood clot non split conditions were taken. On the left and right boundaries of the field to set values of pressure. It was assumed vertical components of velocity are zero at the inlet, free conditions were given on the output of the boundary.

3   Numerical Method

For the numerical solution of the problem used the method of splitting into physical parameters. To solve the system used the approximation on the staggered grid. The presence of thrombus counted by adding a special function in the equations of motion [17] that allows you to accurately represent streamlined border area. Unknown special function determined at the numerical step of the solution problem, thus removing the requirement of elastic border. Below is given a numerical algorithm to determine the dynamics of blood flow.

1. The intermediate speeds $\tilde{u}$, $\tilde{v}$ were determined when $f_u = 0$, $f_v = 0$ on the entire area (outside of the thrombus) explicitly:

$$\frac{\tilde{u} - u^n}{\Delta \tau} = -u^n \frac{\partial u^u}{\partial x} - v^n \frac{\partial u^n}{\partial y} - \frac{\partial p^n}{\partial x} + \frac{1}{Re}\left(\frac{\partial^2 u^n}{\partial x^2} + \frac{\partial^2 u^n}{\partial y^2}\right)$$

$$\frac{\tilde{v} - v^n}{\Delta \tau} = -u^n \frac{\partial v^u}{\partial x} - v^n \frac{\partial v^n}{\partial y} - \frac{\partial p^n}{\partial y} + \frac{1}{Re}\left(\frac{\partial^2 v^n}{\partial x^2} + \frac{\partial^2 v^n}{\partial y^2}\right)$$

2. Then f_u, f_v were determined:

$$f_u = \frac{\tilde{U} - u^n}{\Delta \tau} - u^n \frac{\partial u^u}{\partial x} - v^n \frac{\partial u^n}{\partial y} + \frac{\partial p^n}{\partial x} - \frac{1}{Re}\left(\frac{\partial^2 u^n}{\partial x^2} + \frac{\partial^2 u^n}{\partial y^2}\right)$$

$$f_v = \frac{\tilde{V} - v^n}{\Delta \tau} - u^n \frac{\partial v^u}{\partial x} - v^n \frac{\partial v^n}{\partial y} + \frac{\partial p^n}{\partial y} - \frac{1}{Re}\left(\frac{\partial^2 v^n}{\partial x^2} + \frac{\partial^2 v^n}{\partial y^2}\right)$$

where $\tilde{U}$, $\tilde{V}$−speeds, which are determined by interpolation.

3. Then $\hat{u}$, $\hat{v}$ were determined with f_u, f_v

$$\frac{\hat{u} - u^n}{\Delta \tau} = -u^n \frac{\partial u^u}{\partial x} - v^n \frac{\partial u^n}{\partial y} - \frac{\partial p^n}{\partial x} + \frac{1}{Re}\left(\frac{\partial^2 \hat{u}}{\partial x^2} + \frac{\partial^2 \hat{u}}{\partial y^2}\right) + f_u$$

$$\frac{\hat{v} - v^n}{\Delta \tau} = -u^n \frac{\partial v^u}{\partial x} - v^n \frac{\partial v^n}{\partial y} - \frac{\partial p^n}{\partial y} + \frac{1}{Re}\left(\frac{\partial^2 \hat{v}}{\partial x^2} + \frac{\partial^2 v^n}{\partial y^2}\right) + f_v$$

4. After definition $\hat{u}$, $\hat{v}$, $q = \frac{1}{\Delta x \Delta y}(-\hat{u}\Delta y - \hat{v}\Delta x)$ is determined for cell containing virtual boundaries, that means $f_u \neq 0$, $f_v \neq 0$, $q = 0$ in the fluid, outside of body.

5. Then the equation for quazipressure is solving:

$$\frac{\partial^2 \varphi^{n+1}}{\partial x^2} + \frac{\partial^2 \varphi^{n+1}}{\partial y^2} = \frac{1}{\Delta t}\left(\frac{\partial \hat{u}}{\partial x} + \frac{\partial \hat{v}}{\partial y} - q\right)$$

6. Final values of velocity have form:

$$u^{n+1} = \hat{u} - \Delta t\left(\frac{\partial \varphi^{n+1}}{\partial x}\right)$$

$$v^{n+1} = \hat{v} - \Delta t\left(\frac{\partial \varphi^{n+1}}{\partial y}\right)$$

7. Final field of pressure is defined:

$$p^{n+1} = p^n + \varphi^{n+1} - \frac{\Delta t}{Re}\left(\frac{\partial^2 \varphi^{n+1}}{\partial x^2} + \frac{\partial^2 \varphi^{n+1}}{\partial y^2}\right).$$

Linear and bilinear interpolation were used to improving the order of approximation of the dynamic characteristics on thrombus [18].

4   The Numerical Results

Development of the thrombus is presented in Fig. 1. Initiation of blood clotting due to a local increase in the activator concentration is accompanied by the formation of a blood clot, which displaces the blood flow from the area adjacent to the site of injury. Formation of localized thrombus is determined by the interaction between the activator and an inhibitor and also hydrodynamic flow [14]. In the case of low flow velocities, the wave of coagulation activation is damped by wave of inhibitor and thrombus growth is stopped. Thrombus covers up more one-third of the transverse dimension of the vessel as in [16]. The calculations were performed for $Re = 0.01$, $Pe = 1$ on the grid with size 81x81 and $\delta t = 0.0015$, $\delta x = 0.0125$, $\delta y = 0.00125$ at time $t = 0.45 - (a)$, $t = 1.2 - (b)$, $t = 1.8 - (c)$.

Fig. 2 shows the demolition of clot downstream. The physical cause of that development of thrombosis is the failure of inhibitor to reduce coagulation until threshold value in the case of high speed. The results were obtained with the same parameter values, only $Re = 1$. The increase in speed leads to a qualitative change in the nature of a blood clot. Secondary blood clot appears for away from the injury site in vascular channel. This can be seen from Fig. 3, which shows the results of numerical calculations for $Re = 1$, $Pe = 24$. The complex topological structure of blood clots is the result of nonlinear interaction between activators, inhibitors and a flow in a changing geometry of the area under consideration. Reaction of polymerization provides thrombus growth. The process of thrombus growth affects to the characteristics of the surrounding blood. Thrombus formation in the bloodstream can be activated by a change of parametric of the problem.

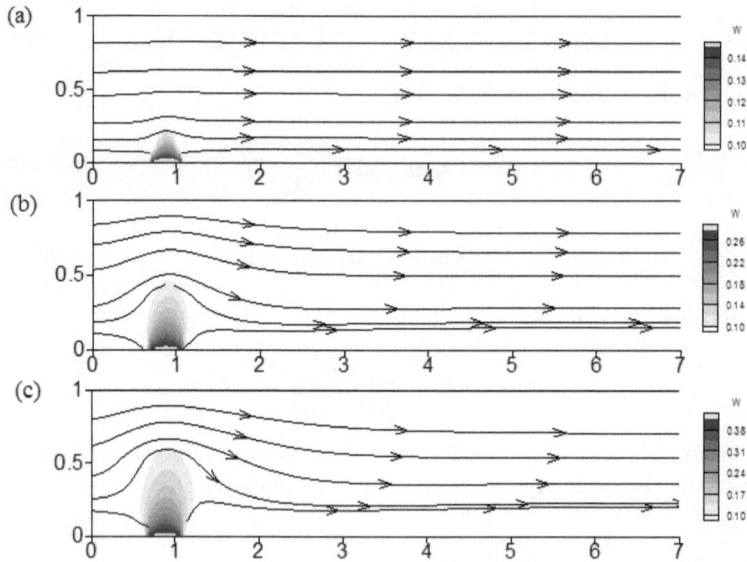

Fig. 1. Thrombus formation at low flow rates: $t = 0.45 - (a)$, $t = 1.2 - (b)$, $t = 1.8 - (c)$, $Re = 0.01$, $Pe = 10$.

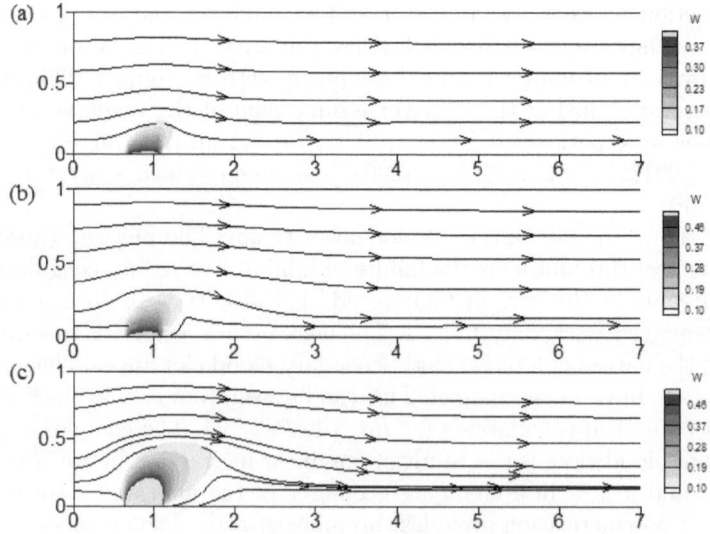

Fig. 2. Thrombus development. $Re = 1$, $Pe = 10$, $t = 0.45 - (a)$, $t = 1.2 - (b)$, $t = 1.8 - (c)$.

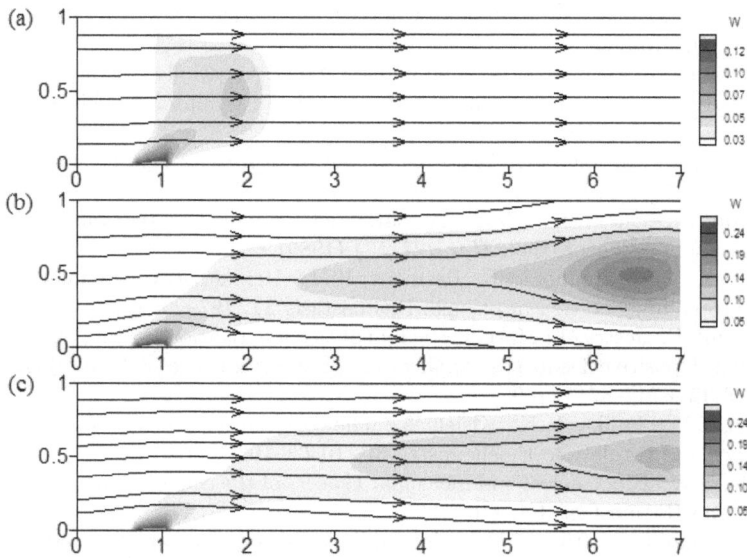

Fig. 3. Thrombus development. $Re = 1$, $Pe = 24$, $t = 0.45 - (a)$, $t = 1.2 - (b)$, $t = 1.8 - (c)$.

5 Conclusion

The aim of this paper was to construct an efficient numerical algorithm for calculating flows in geometrically complex areas. Numerical results obtained by the example of the process of thrombosis, which are qualitatively consistent with the results of other authors [16]. Since the model used is based on a number of assumptions, the most complex development of thrombosis remain outside consideration. The proposed numerical algorithm based on the immersed boundary method increases the accuracy of boundary conditions on the obstacles. Further modification of the proposed algorithm will allow more detailed study of the process of thrombus formation in vessels of complex geometry, including a region with moving boundaries.

References

1. Belotserkovsky, O.M., Kholodov, A.S. (Rep. eds.): Komputernye modeli i progress mediziny. Nauka, Moscow (2001) (in Russian)
2. Belotserkovsky, O.M.: Chislennoe modelirovanie v mexanike sploshnyh sred. Phizmatlit, Moscow (1994). (in Russian)
3. Gosfa, K.D., Hunter, P.J., Pogers, J.M., Gussione, G.M., Waldmen, L.K., McCulloch, A.D.: Three-dimensional limite elements method for large elastic deformations of ventricular myocardium. J. Biomech. **118**, 452–463 (1996)
4. Remizov, A.N.: Medicinskaya I biologicheskaya phizika. Vysshaya shkola, Moscow (1987). (in Russian)

5. Begun, P.S., Afonin, P.N.: Modelirovanie v biomexanike. Vysshaya shkola, Moscow (2004). (in Russian)
6. Evdokimov, A.V., Kholodov, A.S.: Quasistationary spatially distributed model of a closed circulation of the human body. In: Proc. Computer Models and Medical Progress. Science, Moskwa (2001)
7. Davie, E.W., Ratnoff, O.D.: Waterfall sequence for intrinsic blood clotting. Science. **145**, 1310–1312 (1964)
8. Khanin, M.A., Semenov, V.V.: A mathematical model of the kinetics of blood coagulation. J. Theor. Biol. **136**, 127–134 (1989)
9. Willems, G.M., Lindhout, T., Hermens, W.T., Hemker, H.C.: Simulation model for thrombin generation in plasma. Haemostasis **21**, 197–207 (1991)
10. Beltrami, E., Jestry, J.: Mathematical analysis of activation thresholds in enzyme-catalyzed positive feedbacks: application to the feedbacks of blood coagulation. PNAS **92**, 8744–8748 (1995)
11. Davie, E.W., Fujikawa, K., Kisiel, W.: The coagulation cascade: initiation, maintenance and regulation. Biochemistry **30**, 10363–10370 (1991)
12. Davie, E.W.: Biochemical and molecular aspects of the coagulation cascade. Trombosis and Haemostasis **84**, 1–6 (1995)
13. Basmadjian, D., Sefton, M.V., Baldwin, S.A.: Coagulation on biomaterials in flowing blood: some theoretical considerations: Review. Biomaterials **18**, 1511–1522 (1997)
14. Ataullakhanov, F.I., Guria, G.T., Sarbash, V.I., Volkova, R.I.: Spatio-temporal dynamics of clotting and pattern formation in human blood. Biochimca et Biophysica Acta **14**(25), 453–468 (1998)
15. Peskin, C.S.: Numerical analysis of blood flow in the heart. J. Comput. Phys. **25**, 220–252 (1977)
16. Chulichkov, A.L., Nikolaev, A.V., Lobanov, A.I., Guriya, G.T.: Porogovaya aktivasia cvertyvaniya krovi i rost tromba v usloviyax krovotoka. Matematicheskoe modelirovanie **12**(3), 75–96 (2000). (in Russian)
17. Peskin, C.S., McQueen, D.M.: A three-dimensional computational method for blood flow in the heart: (I) immersed elastic fibers in a viscous incompressible fluid. J. Comput. Phys. **81**, 372–405 (1989)
18. Kim, J., Kim, D., Choi, H.: An immersed-boundary finite-volume method for simulations of flow in complex geometries. Journal of Computational Physics **171**, 132–150 (2001)

Modeling the Impact of Relief Boundaries in Solving the Direct Problem of Direct Current Electrical Sounding

Balgaisha Mukanova and Tolkyn Mirgalikyzy[(✉)]

L.N. Gumilyov Eurasian National University, Satpayev str. 2,
010008 Astana, Kazakhstan
mbsha01@gmail.com, m_t85@mail.ru

Abstract. In this study we examined the numerical methods of solving the direct problem of electrical sounding with direct current for a layered model with complex relief contact boundaries. The solution was obtained by the method of integral equations. The system of integral equations for the solution of the direct problem of electrical sounding with direct current for a layered relief medium was established. Numerical simulation of the field for two-layered medium with various shapes of relief contact boundaries was conducted. We obtained the density of distribution of secondary sources on contact boundaries.

Keywords: The direct problem of electrical sounding · Layered relief medium · System of integral equations

1 Introduction

The theory of electrical sounding are usually designed for a medium with a flat surface. However, complex structures of surface relief frequently come up in practice. Accordingly the task of studying the impact of various forms of ground surface on the results of geophysical investigations raises. Today some basic methods of solving the direct problem of electrical sounding with direct current are developed: method of integral equations [1],[2],[3] finite difference method [4], finite element method [5], boundary element method [6].

Cases of ground surface relief of the Earth were not considered in investigations. Or the cases were not brought to systematical numeral modeling. Currently available methods of relief form corrections have approximate pattern. In this study, for the calculation of fields in layered relief medium, we use the method of integral equations, well-established when performing modeling in resistivity method [7],[8],[9].

2 Mathematical Model

Let us consider the case of medium with surface relief Γ_0 and piecewise constant distribution of the electrical conductivity $\sigma(M)$. Let the medium be divided into

© Springer International Publishing Switzerland 2015
N. Danaev et al. (Eds.): CITech 2015, CCIS 549, pp. 117–123, 2015.
DOI: 10.1007/978-3-319-25058-8_12

N areas with constant conductivity $\sigma_1, \sigma_2, ..., \sigma_N$. Let the boundaries between the areas form two-dimensional piecewise smooth surfaces, which involve defined normal lines almost everywhere. Let σ_0 be the conductivity of the medium, where the source electrode is placed on the surface, and Γ_0 is its ground surface. We assume that the source electrode does not fall on one of the boundaries between the contacting media. Let $\Gamma_1, \Gamma_2, ..., \Gamma_k$ be areas of surfaces of contacting media with different conductivities, and surface media in contact with air. It will be denoted by the values σ_i^+, σ_i^- of the conductivity of the surface Γ_i from different angles for media that have a common internal border Γ_i. We introduce new unknowns functions $q_1, q_2, ..., q_k$, defined on these parts of the boundaries with the meaning of the surface density of the secondary charges distributed in these boundary areas.

We seek the field potential in the form of a sum of simple layer potentials defined on these parts of the boundaries and the potential of a point source in a half-space:

$$U(P) = U_0(P) + \frac{I}{4\pi\sigma_0} \sum_{k=0}^{K} u_k(P) = \frac{I}{2\pi\sigma_0 |AP|} + \frac{I}{4\pi\sigma_0} \sum_{k=0}^{K} \int_{\Gamma_k} \frac{q_k(M)}{|PM|} d\Gamma(M) \tag{1}$$

Obviously, that way a particular function satisfies the Laplace equation in the areas of consistency σ. We require that the function $U(P)$ satisfy the boundary conditions:

$$\begin{cases} \Delta u = 0 \\ \sigma \frac{\partial u}{\partial n}\Big|_{\Gamma} = -\sigma \frac{\partial U_0}{\partial n}\Big|_{\Gamma} + \frac{I}{2\pi} \delta(\overline{r} - \overline{OA}) \\ u(\infty) = 0 \end{cases} \tag{2}$$

We provide additional conditions on the inner contact boundaries. These conditions mean a continuous flow of charge through the contact boundaries and can be written as:

$$\sigma_i^+ \left(\frac{\partial U}{\partial n}\right)_+ = \sigma_i^- \left(\frac{\partial U}{\partial n}\right)_- \tag{3}$$

We rewrite (3) in terms of the functions $u_i(P) = \int_{\Gamma_i} \frac{q_k(M)d\Gamma(M)}{|PM|}$, considering that the derivatives of the potential of a simple layer with density $q_i(P)$ in the i-th boundary have discontinuity at this contact boundary, and derivatives of the potential from other sources are continuous:

$$\sigma_i^+ \left(\frac{\partial u_i}{\partial n}\right)_+ = \sigma_i^- \left(\frac{\partial u_i}{\partial n}\right)_- - (\sigma_i^+ - \sigma_i^-) \left(\frac{4\pi\sigma_0}{I} \frac{\partial U_0}{\partial n} + \sum_{k\neq i}^{K} \frac{\partial u_k}{\partial n}(P)\right), \tag{4}$$

$$i = 1, 2, ..., K$$

The boundary condition (2) at all points, except at the point A, can be rewritten as:

$$\frac{I}{4\pi\sigma_0} \sum_{k=0}^{K} \frac{\partial u_k}{\partial n}(P)\Big|_{\Gamma_0} + \frac{\partial U_0}{\partial n}\Big|_{\Gamma_0} = 0 \tag{5}$$

Thus, we have exactly as many relations as unknown functions q_k – densities of induced charges on parts of contact boundaries. To formulate these relations in the form of a system of integral equations we write discontinuity property of the normal derivative of simple layer for internal contact boundaries:

$$\left(\frac{\partial u_i}{\partial n}\right)_+ = -2\pi q_i\left(P\right) + \int_{\Gamma_i} q_i\left(M\right) \frac{\cos \psi_{PM}}{|PM|^2} d\Gamma_i\left(M\right)$$

$$\left(\frac{\partial u_i}{\partial n}\right)_- = 2\pi q_i\left(P\right) + \int_{\Gamma_i} q_i\left(M\right) \frac{\cos \psi_{PM}}{|PM|^2} d\Gamma_i\left(M\right)$$

(6)

3   Derivation of the System of Integral Equations and Numerical Results

We derive a system of integral equations for the densities of a simple layer and substitute the formula (6) into the boundary conditions (4):

$$\sigma_i^+ \left(-2\pi q_i\left(P\right) + \int_{\Gamma_i} q_i\left(M\right) \frac{\cos \psi_{PM}}{|PM|^2} d\Gamma_i\left(M\right)\right) = $$

(7)

$$\sigma_i^- \left(2\pi q_i\left(P\right) + \int_{\Gamma_i} q_i\left(M\right) \frac{\cos \psi_{PM}}{|PM|^2} d\Gamma_i\left(M\right)\right) - $$

$$- \left(\sigma_i^+ - \sigma_i^-\right) \left(\frac{4\pi\sigma_0}{I} \frac{\partial U_0}{\partial n} + \sum_{k\neq i}^{K} \frac{\partial u_k}{\partial n}\left(P\right)\right)$$

After obvious simplifications, we obtain:

$$q_i\left(P\right) = \frac{\lambda}{2\pi} \int_{\Gamma_i} q_i\left(M\right) \frac{\cos \psi_{PM}}{|PM|^2} d\Gamma_i\left(M\right) + \frac{\lambda}{2\pi} \left(\frac{4\pi\sigma_0}{I} \frac{\partial U_0}{\partial n} + \sum_{k\neq i}^{K} \frac{\partial u_k}{\partial n}\left(P\right)\right), \quad (8)$$

$$i = \overline{1, K}$$

where $\lambda = \frac{\left(\sigma_i^+ - \sigma_i^-\right)}{\left(\sigma_i^+ + \sigma_i^-\right)}$ is the reflectivity factor at the boundaries of two contacting medium. Substituting in (8) function of the potential generated by charge on other boundaries, we get:

$$q_i\left(P\right) = \frac{\lambda}{2\pi} \int_{\Gamma_i} q_i\left(M\right) \frac{\cos \psi_{PM}}{|PM|^2} d\Gamma_i\left(M\right) + \frac{\lambda}{\pi} \frac{\partial}{\partial n} \frac{1}{|AP|} + $$

(9)

$$\frac{\lambda}{2\pi} \sum_{k\neq i}^{K} \int_{\Gamma_k} q_k\left(M\right) \frac{\cos \psi_{PM}}{|PM|^2} d\Gamma\left(M\right)$$

For the density of a simple layer on the outer boundary Γ_0 the integral equation is obtained from the boundary condition (5). In terms of density of a simple layer, the condition (5) after elementary simplifications can be written as:

$$\left(\frac{\partial u_0}{\partial n}\right)_+ = \frac{\partial}{\partial n_+}\left(\int_{\Gamma_0} q_0\left(M\right)\frac{\cos\psi_{PM}}{|PM|^2}d\Gamma\left(M\right)\right)\Big|_{\Gamma_0} = -\frac{4\pi\sigma_0}{I}\frac{\partial U_0}{\partial n}\Big|_{\Gamma_0} - \sum_{j=1}^{K}\frac{\partial u_j}{\partial n}\Big|_{\Gamma_j} \tag{10}$$

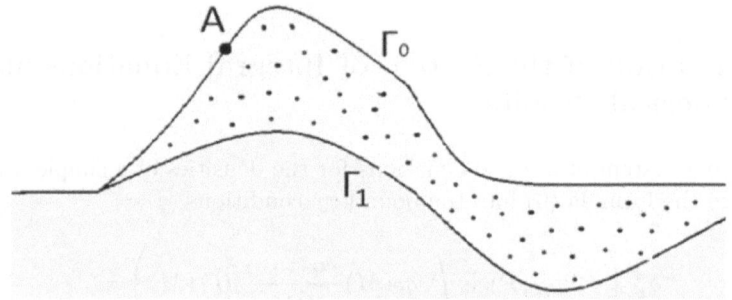

Fig. 1. Two-layered medium with relief boundaries

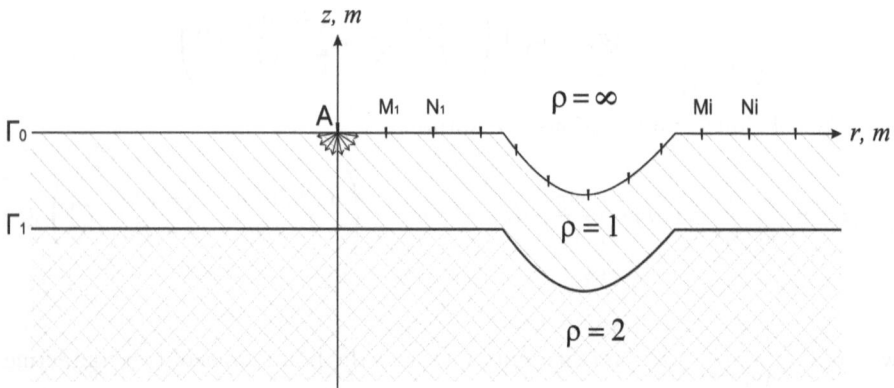

Fig. 2. The model of the medium

With regard to the expression of the normal derivative for the outer side of surface we obtain

$$\left(\frac{\partial u_0}{\partial n}\right)_+ = -2\pi q_0\left(P\right) + \int_{\Gamma} q_0\left(M\right)\frac{\cos\psi_{PM}}{|PM|^2}d\Gamma\left(M\right) \tag{11}$$

a)

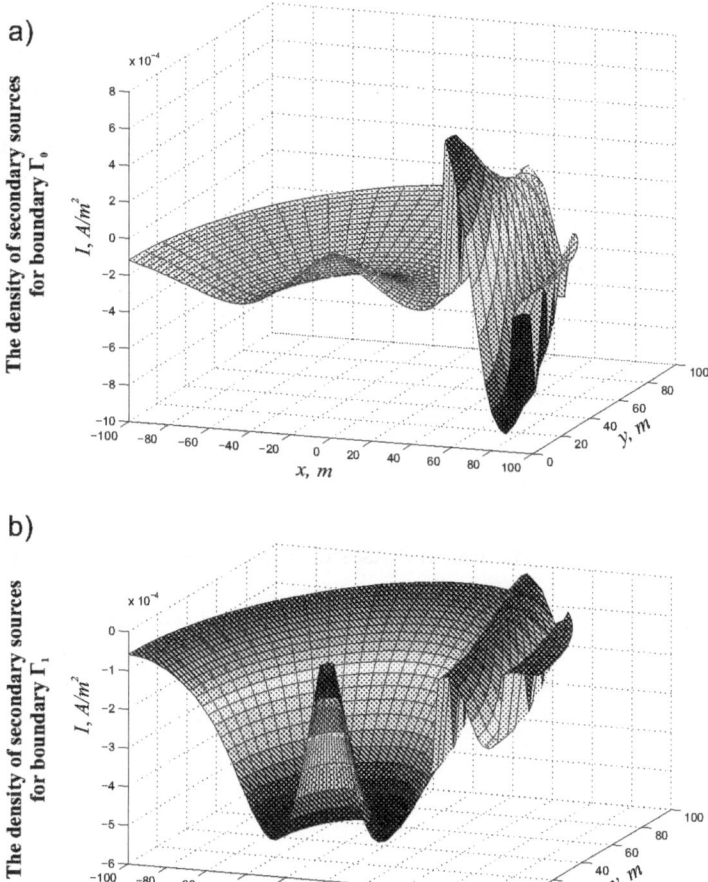

b)

Fig. 3. The densities of secondary source on boundaries: a) on boundary Γ_0; b) on boundary Γ_1

Put (10) in (9):

$$-2\pi q_0\left(P\right) + \int_\Gamma q_0\left(M\right)\frac{\cos\psi_{PM}}{\left|PM\right|^2}d\Gamma\left(M\right) = -\frac{4\pi\sigma_0}{I}\frac{\partial U_0}{\partial n}\bigg|_{\Gamma_0} - \sum_{j=1}^{K}\frac{\partial u_j}{\partial n}\bigg|_{\Gamma_0} =$$

$$= -\frac{4\pi\sigma_0}{I}\frac{\partial U_0}{\partial n}\bigg|_{\Gamma_0} - \sum_{j=1}^{K}\int_{\Gamma_j} q_j\left(M\right)\frac{\cos\psi_{PM}}{\left|PM\right|^2}d\Gamma_j\left(M\right) \quad (12)$$

or

$$q_0\left(P\right) - \frac{1}{2\pi}\sum_{j=0}^{K}\int_{\Gamma_j} q_j\left(M\right)\frac{\cos\psi_{PM}}{\left|PM\right|^2}d\Gamma_j\left(M\right) = \frac{1}{\pi}\frac{\partial}{\partial n}\frac{1}{\left|AP\right|}\bigg|_{\Gamma_0} \quad (13)$$

Equations (9), (13) form a system of equations for the densities of a simple layer defined on the contact surfaces of medium and a ground surface.

Here are some examples of systems of integral equations for different models of medium.

For example, for the medium model, shown in figure 1, we can write the following system of integral equations:

$$
\begin{aligned}
q_0\left(P\right) - \frac{1}{2\pi} \int_{\Gamma_0} q_0\left(M\right) \frac{\cos \psi_{PM}}{\left|PM\right|^2} d\Gamma_0\left(M\right) - \\
\frac{1}{2\pi} \int_{\Gamma_1} q_1\left(M\right) \frac{\cos \psi_{PM}}{\left|PM\right|^2} d\Gamma_1\left(M\right) = \frac{1}{\pi} \frac{\partial}{\partial n} \frac{1}{\left|AP\right|}, P \in \Gamma_0 \\
q_1\left(P\right) - \frac{\lambda}{2\pi} \int_{\Gamma_1} q_1\left(M\right) \frac{\cos \psi_{PM}}{\left|PM\right|^2} d\Gamma_1\left(M\right) - \\
\frac{\lambda}{2\pi} \int_{\Gamma_0} q_0\left(M\right) \frac{\cos \psi_{PM}}{\left|PM\right|^2} d\Gamma_0\left(M\right) = \frac{\lambda}{\pi} \frac{\partial}{\partial n} \frac{1}{\left|AP\right|}, P \in \Gamma_1
\end{aligned}
\tag{14}
$$

Similarly, the equations for the cases of buried relief or local inclusion are written out.

In figure 3 the density of secondary sources for boundaries Γ_0 and Γ_1 are plotted. The solution obtained by solving the system of integral equations. Calculations were made for a three-electrode Schlumberger array (AMN) for two-layered relief medium with resistivities $\rho_1 = 1 \ \Omega m$ and $\rho_2 = 2 \ \Omega m$. The model of the medium is shown in figure 2.

4   Conclusion

A system of integral equations for calculating the density of the secondary currents at the boundaries of contacting media in case of multiple layered relief media was established. The proposed method allows to calculate the field of the point source on the ground surface, which has a relief shape. Numerical examples of the calculation of fields of current densities of secondary sources were shown.

References

1. Alpine, L.: Field source in the theory of electrical prospecting. Applied Geophysics **3**, 56–200 (1947). (in Russian)
2. Hvozdara, M.: Electric and magnetic field of a stationary current in a stratified medium with a three-dimensional conductivity inhomogeneity. Srudia Geophysica et Geodaetica **26**, 59–84 (1983)
3. Khmelevskoy, V., Shevnin, V.: Electrical sounding of geological medium. Moscow (1999)
4. Loke, M.: 2-D and 3-D electrical imaging surveys. The tutorial (2004)
5. Gunther, T., Rucker, C.: Boundless Electrical Resistivity Tomography. BERT 2 - the user tutorial, version 2.0 (2013)

6. Xu, S., Zhao, S., Ni, Y.: A boundary element method for 2-D DC resistivity modeling with a point current source. Geophysics **63**, 399–404 (1998)
7. Orunkhanov, M., Mukanova, B., Sarbasova, B.: Convergence of an integral equation on geoelectric sounding problem above a local patch. Computational Technologies **9**(6), 68–72 (2004). (in Russian), View at Google Scholar
8. Orunkhanov, M., Mukanova, B.: The integral equations method in problems of electrical sounding. In: Advances in High Performance Computing and Computational Sciences, pp. 15–21. Springer, Heidelberg (2006)
9. Orunkhanov, M., Mukanova, B., Sarbasova, B.: Convergence of the method of integral equations for quasi three-dimensional problem of electrical sounding. In: Computational Science and High Performance Computing II, pp. 175–180. Springer, Heidelberg (2006)

Quantum Computing and Its Potential for Turbulence Simulations

S. Sammak[1], A.G. Nouri[1], N. Ansari[1,2], and P. Givi[1(✉)]

[1] Mechanical Engineering and Materials Science, University of Pittsburgh,
Pittsburgh, PA 15261, USA
[2] ANSYS Inc., 275 Technology Dr., Canonsburg, PA 15317, USA
{shs159,arn36,naa56,pgivi}@pitt.edu

Abstract. A tutorial is provided of quantum computing (QC) and the
way it has made significant speed-up in various simulations. A review
will also be provided of the large eddy simulation (LES) of turbulent
flows via the stochastic filtered density function (FDF) methodology. The
potentials of the quantum speed-up in FDF simulation via QC appear
to be significant. This can results to a revolutionary means by which
turbulence simulations can be conducted in future.

1 Introduction

Quantum computation (QC) has undergone rapid development, both experimentally and theoretically, in recent years [1]. Used in appropriate ways, quantum
mechanics can provide powerful resources for solving certain classes of problems, achieving speedups not available to classical computers. The best known
examples are Shor's algorithm for factorization of integers [2], and Grover's algorithm for unstructured search problems [3]. The gain in efficiency can either be
exponential (*i.e.*, a problem where the solution time on a classical computer
scales exponentially in the size of the problem N can have a solution time
that scales polynomially in that size on a quantum computer), or polynomial
(*i.e.*, the problem scales polynomially with N on a classical computer, and with
a smaller power of N on a quantum computer) [1]. In either case, for the solution
of large-scale problems, quantum computers represent a potentially transformational new paradigm in computing.

Within the past decade much progress has also been made in experimental realizations of quantum computing hardware. Many architectures have been
proposed based on a variety of physical hardware. On a small scale, quantum
information has been stored and manipulated in superconducting quantum bits
(qubits) [4,5], trapped ions [6,7], electron spins [8–11], nuclear spins in the liquid or solid state [12], and other systems. On the theoretical side, new quantum
algorithms have recently been found, exhibiting significant polynomial speedups
on quantum computers for solution of sparse linear equations or differential
equations [13,14], quantum Monte Carlo problems [15], and classical simulated
annealing problems [16].

© Springer International Publishing Switzerland 2015
N. Danaev et al. (Eds.): CITech 2015, CCIS 549, pp. 124–132, 2015.
DOI: 10.1007/978-3-319-25058-8_13

It is speculated that QC can be a very useful tool for simulation of turbulent reacting flows. These flows are of significant interest to many industries as well as several sectors of the government. A possible means of achieving this is to search for quantum algorithms which are capable of solving stochastic differential equations (SDEs) which are central to the classical large eddy simulation (LES) methods. The optimal means of capturing the detailed physics of such flows via LES utilizes the density function (FDF) methodology [17]. The FDF provides the most comprehensive form of accounting for the subgrid scale (SGS) quantities and it can be cast in the form of the Diffusion equation [17]:

$$d\mathbf{X} = \mathbf{A}dt + Bd\mathbf{W} , \tag{1}$$

where $\mathbf{X}$ is a vector specifying all of the fluid and thermodynamic variables associated with the flow, vector $\mathbf{A}$ and matrix B are the drift and diffusion coefficients, respectively, components of which are specific to a given FDF model, t is time and $\mathbf{W}$ is the set of independent Weiner processes. For LES to be practical it must be conducted in a computationally efficient manner, especially if it is employed for prediction of complex flows.

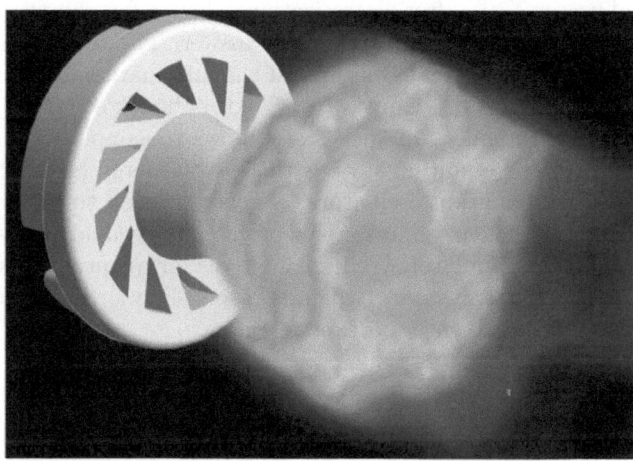

Fig. 1. Classical FDF Prediction of CO Mass Fraction in a Bluff-Body Reactor [21].

"Classical" computation of the FDF has experienced tremendous progress within the past decade; see *e.g.* Refs. [18,19]. As a results, it is now possible to conduct simulation of some of the most complex reacting flows. As an example, Fig. 1 shows LES/FDF prediction of a realistic chemical reactor [20]. The FDF simulated results as shown are within 95 % agreement with experimental measurements. However, some of these classical simulations can take of order of *several months* [21]!

2   QC for SDE Simulation

We speculate that quantum algorithms can be developed for efficient simulation of SDEs using two different methods. The first method is "algebraic" and computes arbitrary entries of the vector $\mathbf{X}$ in Eq. (1) using fast quantum algorithms for matrix multiplication. The second method prepares a quantum state such that, after a simple measurement, it samples from a probability distribution as determined by general Wiener processes. This method is "physical" in that the quantum states are prepared by quantum adiabatic evolutions of Hamiltonians that model the quantum harmonic oscillator and could be implemented by quantum simulators. In both methods, techniques from quantum walks developed in Ref. [16] to speedup conventional Monte-Carlo can be used. For that it is needed to investigate different reformulations of the FDF algorithm that are more amenable to implementation on a quantum computer.

Since the FDF is essentially modelled via a set of SDEs, it may be actually possible to employ QC for its simulations. This can be potentially achieved in two different ways. In the first case a simulation is aimed to output an arbitrary entry of the vector $\mathbf{X}(t)$, that satisfies a type of Eq. (1), with specified accuracy. This method works because $\mathbf{X}(t)$ can be sometimes written as a matrix product acting on some initial vector $\mathbf{X}(0)$, after discretization. Then, a fast quantum algorithm for matrix multiplication would provide the answer. The second method aims to prepare a quantum state that contains all the information about the evolved probability distribution as determined by a Fokker-Planck equation like Eq. (5) below. A simple measurement in such state allows to sample with exactly the desired probability. The quantum state is the lowest-energy (ground) state of a system of perturbed quantum oscillators. A generalization to the discrete case would provide quantum algorithms for this problem that could be readily implemented on a quantum simulator. Well-known tools for quantum speedups, such as the quantum Fourier transform (which is responsible for fast factoring as in Shor's algorithm) will play an important role here.

3   Turbulence Formulation via FDF

As indicated previously, the idea of employing QC for turbulence (including turbulent combustion) simulation appears promising because the essential means of enacting FDF is via modeled SDEs. These SDEs describe all of the basic transport variables and account for couplings of turbulence, exothermicity, variable density, and also differential diffusion. The primary transport variables in FDF formulation are the density $\rho(\mathbf{x}, t)$, the velocity vector $u_i(\mathbf{x}, t)$ $(i = 1, 2, 3)$, the pressure $p(\mathbf{x}, t)$, the internal energy $e(\mathbf{x}, t)$ and the species mass fractions ϕ_α $(\alpha = 1, \ldots N_s)$. The equations which govern the transport of these variables in space (x_i) $(i = 1, 2, 3)$ and time (t) are the continuity, momentum, energy and the scalar transport, all coupled through the equation of state [22].

$$\frac{\partial \rho}{\partial t} + \frac{\partial \rho u_j}{\partial x_j} = 0 \tag{2a}$$

$$\frac{\partial \rho u_i}{\partial t} + \frac{\partial \rho u_j u_i}{\partial x_j} = -\frac{\partial p}{\partial x_i} + \frac{\partial \tau_{ji}}{\partial x_j} \tag{2b}$$

$$\frac{\partial \rho e}{\partial t} + \frac{\partial \rho u_j e}{\partial x_j} = -\frac{\partial q_j}{\partial x_j} + \sigma_{ij}\frac{\partial u_i}{\partial x_j} \tag{2c}$$

$$\frac{\partial \rho \phi_\alpha}{\partial t} + \frac{\partial \rho u_j \phi_\alpha}{\partial x_j} = -\frac{\partial J_j^\alpha}{\partial x_j} + S(\Phi) \tag{2d}$$

$$p = \rho \frac{R^0}{W} T \tag{2e}$$

where R^0 denotes the universal gas constant and W is the mean molecular weight of the mixture. T denote the temperature, e is the internal energy $\gamma = \frac{c_p}{c_v}$ is the specific heat ratio, and $S(\Phi)$ denotes the chemical source term. The viscous stress tensor τ_{ij}, the energy flux q_j, the species α diffusive mass flux vector J_j^α and σ_{ij} tensor are represented by

$$\sigma_{ij} = \tau_{ij} - p\delta_{ij}, \quad \tau_{ij} = \mu\left(\frac{\partial u_i}{\partial x_j} + \frac{\partial u_j}{\partial x_i} - \frac{2}{3}\frac{\partial u_k}{\partial x_k}\delta_{ij}\right),$$

$$q_j = -\lambda\frac{\partial T}{\partial x_j}, \quad J_j^\alpha = -\rho\Gamma\frac{\partial \phi_\alpha}{\partial x_j}, \tag{3}$$

where μ is the fluid dynamic viscosity, λ denotes the thermal conductivity and Γ is the mass diffusion coefficient. We assume calorically perfect gas in which the specific heats are constants. Large eddy simulation involves the spatial filtering operation: [23,24] $\langle Q(\mathbf{x},t)\rangle_\ell = \int_{-\infty}^{+\infty} Q(\mathbf{x}',t)G_{\Delta_l}(\mathbf{x}',\mathbf{x})d\mathbf{x}'$, where $G_{\Delta_l}(\mathbf{x}',\mathbf{x})$ denotes a filter function, and $\langle Q(\mathbf{x},t)\rangle_\ell$ denotes the filtered value of the transport variable $Q(\mathbf{x},t)$. The subscript l_1 indicates that $\langle Q(\mathbf{x},t)\rangle_\ell$ is the first level filter value of the variable $Q(\mathbf{x},t)$ [25]. In variable-density flows it is convenient to use the Favre-filtered quantity $\langle Q(\mathbf{x},t)\rangle_L = \langle \rho Q\rangle_\ell / \langle \rho\rangle_\ell$. We consider a filter function that is spatially and temporally invariant and localized, thus: $G_{\Delta_{l_1}}(\mathbf{x}',\mathbf{x}) \equiv G_{\Delta_{l_1}}(\mathbf{x}'-\mathbf{x})$ with the properties $G_{\Delta_{l_1}}(\mathbf{x}) \geq 0$, $\int_{-\infty}^{+\infty} G_{\Delta_{l_1}}(\mathbf{x})d\mathbf{x} = 1$.

The formal FDF is defined by $F_L(\boldsymbol{v}, \boldsymbol{\psi}, \boldsymbol{\Theta}, \boldsymbol{\eta}, \mathbf{x}; t)$ where, $\boldsymbol{v}$, $\boldsymbol{\psi}$, $\boldsymbol{\Theta}$ and $\boldsymbol{\eta}$ are the velocity vector, the scalar array, the sensible internal energy and pressure in the sample space, respectively. The function F has all of the properties of a probability density function in that the filtered value of any function of the velocity and/or scalar variables is obtained by its integration over the sample spaces:

$$\langle \rho(\mathbf{x},t)\rangle_\ell \langle Q(\mathbf{x},t)\rangle_L = \int_{-\infty}^{+\infty}\int_{-\infty}^{+\infty} Q(\boldsymbol{v},\boldsymbol{\psi},\boldsymbol{\Theta},\boldsymbol{\eta})F_L(\boldsymbol{v},\boldsymbol{\psi},\boldsymbol{\Theta},\boldsymbol{\eta},\mathbf{x};t)d\boldsymbol{v}d\boldsymbol{\psi}d\boldsymbol{\Theta}d\boldsymbol{\eta}. \tag{4}$$

However, the transport equation for F_L is not in a closed form and must be modelled. For this, as indicted earlier, we consider the general diffusion process,[26,27] given by the system of SDEs. The modeling of the SDEs must be in such a way that is amenable to QC. The starting point will be our simplified Langevin model (SLM) and linear mean-square estimation (LMSE) [28] coupled with an equation of state and obeying the first law of thermodynamics. With construction of the SDEs, the corresponding Fokker-Planck equation [29] will essentially be the modelled FDF transport equation. Our proposed model is under construction and is of the form:

$$\frac{\partial F_L}{\partial t} + \frac{\partial v_i F_L}{\partial x_i} = \frac{1}{\langle \rho \rangle_\ell} \frac{\partial \langle p \rangle_\ell}{\partial x_i} \frac{\partial F_L}{\partial v_i} - \frac{2}{\langle \rho \rangle_\ell} \frac{\partial}{\partial x_j} \left(\mu \frac{\partial \langle u_i \rangle_L}{\partial x_j} \right) \frac{\partial F_L}{\partial v_i} - \frac{1}{\langle \rho \rangle_\ell} \frac{\partial}{\partial x_j} \left(\mu \frac{\partial \langle u_j \rangle_L}{\partial x_i} \right) \frac{\partial F_L}{\partial v_i}$$

$$+ \frac{2}{3} \frac{1}{\langle \rho \rangle_\ell} \frac{\partial}{\partial x_i} \left(\mu \frac{\partial \langle u_j \rangle_L}{\partial x_j} \right) \frac{\partial F_L}{\partial v_i} - \frac{\partial \left(G_{ij} \left(v_j - \langle u_j \rangle_L \right) F_L \right)}{\partial v_i} + \frac{\partial}{\partial x_i} \left(\mu \frac{\partial (F_L / \langle \rho \rangle_\ell)}{\partial x_i} \right)$$

$$+ \frac{\partial}{\partial x_i} \left(\frac{2\mu}{\langle \rho \rangle_\ell} \frac{\partial \langle u_j \rangle_L}{\partial x_i} \frac{\partial F_L}{\partial v_j} \right) + \frac{\mu}{\langle \rho \rangle_\ell} \frac{\partial \langle u_k \rangle_L}{\partial x_j} \frac{\partial \langle u_i \rangle_L}{\partial x_j} \frac{\partial^2 F_L}{\partial v_k \partial v_i} + \frac{1}{2} C_0 \frac{\epsilon}{\langle \rho \rangle_\ell} \frac{\partial^2 F_L}{\partial v_i \partial v_i}$$

$$+ C_\phi \omega \frac{\partial \left((\psi_\alpha - \langle \phi_\alpha \rangle_L) F_L \right)}{\partial \psi_\alpha} + \frac{C_e \omega}{\gamma} \frac{\partial \left((\theta - \langle e \rangle_L) F_L \right)}{\partial \theta}$$

$$- \epsilon \frac{\gamma - 1}{\gamma} \frac{\partial}{\partial \theta} \left(\frac{\theta}{\eta} F_L \right) - \frac{\gamma - 1}{\gamma} \frac{\partial (\theta A F_L)}{\partial \theta} + \frac{\gamma - 1}{\gamma^2} \frac{\partial (\theta B^2 F_L)}{\partial \theta} - \frac{\partial (\eta A F_L)}{\partial \eta}$$

$$+ \frac{1}{2} \frac{(\gamma - 1)^2}{\gamma^2} \frac{\partial^2 (\theta^2 B^2 F_L)}{\partial \theta \partial \theta} + \frac{\gamma - 1}{\gamma} \frac{\partial^2 (\theta \eta B^2 F_L)}{\partial \theta \partial \eta} + \frac{1}{2} \frac{\partial^2 (\eta^2 B^2 F_L)}{\partial \eta \partial \eta}.$$

$$(5)$$

In Eq. (5) k is the SGS kinetic energy, $\epsilon = \langle \rho \rangle_\ell C_\epsilon \frac{k^{3/2}}{\Delta_L}$ is the SGS dissipation, $\omega = \frac{1}{\langle \rho \rangle_\ell} \frac{\epsilon}{k}$ is the SGS frequency, A, B are the model parameters for the pressure SDE, and

$$G_{ij} = \left[\frac{\Pi_d}{2k \langle \rho \rangle_\ell} - \omega \left(\frac{1}{2} + \frac{3}{4} C_0 \right) \right] \delta_{ij}. \tag{6}$$

The parameters C_0, C_ϕ, C_e, and C_ϵ are model constants and need to be specified [30,31]. The same goes for the pressure dilatation term Π_d [31,32]. The transport equations for all of the SGS moments are readily obtained by integration of this Fokker-Planck equation. This provides a complete statistical description of turbulence. The idea is to find methods that could take advantage of quantum resources in order to speed up these calculations, at least polynomially in the number of variables. Because of the size of the problem typically considered, such a speedup could transform the way these problems are treated in engineering; providing solutions to problems many orders of magnitude faster than are possible with classical computers.

Another challenge associated with FDF is its implementation in *complex* geometries. Structured multi-block grids lack the required flexibility and robustness for handling such geometries. The grid cells may also become too skewed

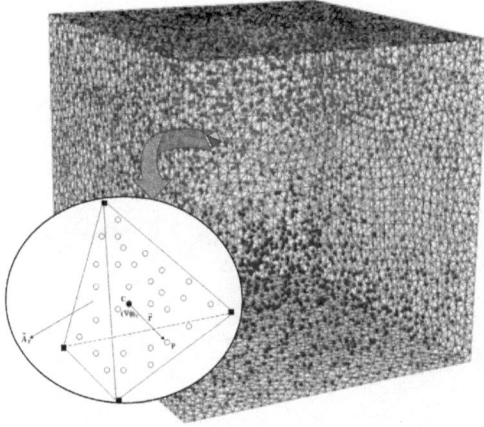

Fig. 2. Monte Carlo Particles on Unstructured Grids.

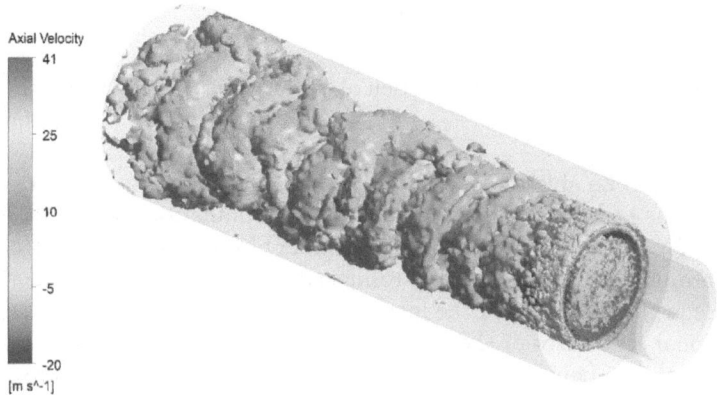

Fig. 3. Classical FDF Simulation of the Sandia/Sydney Burner [36].

and/or twisted, prohibiting efficient simulations. Unstructured grids provide a good solution for the problem of producing grids on complex shapes. Such grids have irregularly distributed nodes and their cells are not required to be of a specific shape. Furthermore, the connectivity of neighboring cells can vary spatially. The SDEs portraying the FDF will be primarily simulated via Monte Carlo methods [33,34], on a domain represented by unstructured grids. As an example, Fig. 2 shows a small fraction of the Monte Carlo particles used on unstructured meshes for the classical simulation results shown in Fig. 1. The essence of a quantum computing implementation would be to speed up these Monte Carlo simulations, either in direct solutions of the stochastic differential equations, or by solving equivalent problems (such as the Fokker-Planck equations).

Fig. 4. Classical FDF Simulation of Taylor-Green Vortex Flow [37].

We expect to be able to conduct LES with over billions grid points and over tens of billions of Monte Carlo particles. These simulations are proposed to be conducted for prototype reactors with variable physical length and time scales. In this case, the effects of the flow residence time and the Damköhler number will be the primary subject of the investigations. In addition, the spatial and the compositional structures of the reacting flow field will be assessed.

At this end, to demonstrate superiority of our classical algorithms, we show some sample results of our most recent FDF simulation of the Sandia/Sydney swirl burner [35]. This configuration is selected as it is one of the most challenging turbulent flames for prediction. Figure 3 shows the contours of the azimuthal velocity field as predicted by our FDF. The simulated results agree with experimental data better than any other classical methods currently available [36]. But the computational time requirements are excessive. As another example, Fig. 4 shows the contour of filtered temperature field for the symbolic Taylor-Green vortex flow as obtained via FDF coupled with a discontinuous Galerkin flow solver [37]. Quantum computation may potentially provide a much more efficient means for such simulations.

Acknowledgments. This work is sponsored by AFOSR under Grant FA9550-12-1-0057, by NSF under Grant CBET-1250171, and by the NSF Extreme Science and Engineering Discovery Environment (XSEDE) under Grants TG-CTS070055N and TG-CTS120015. We are thankful to members of the Center for Simulation and Modeling at the University of Pittsburgh for their help with numerous computational issues.

References

1. Nielsen, M.A., Chuang, I.L.: Quantum computation and quantum information. Cambridge University Press, Cambridge (2010)
2. Shor, P.W.: Polynomial-Time Algorithms for Prime Factorization and Discrete Logarithms on a Quantum Computer. SIAM Rev. **41**(2), 303–332 (1999)
3. Grover, L.K.: A fast quantum mechanical algorithm for database search. In: Proceedings of the Twenty-Eighth Annual ACM Symposium on Theory of Computing, pp. 212–219, New York, NY (1996)
4. DiVincenzo, D.P.: Fault-tolerant Architectures for Superconducting Qubits. Phys. Scripta **2009**(T137), 014020 (2009)
5. Geller, M., Pritchett, E., Sornborger, A., Wilhelm, F.: Quantum computing with superconductors I: architectures. In: Manipulating Quantum Coherence in Solid State Systems, pp. 171–194. Springer, Netherlands (2007)
6. Leibfried, D., Blatt, R., Monroe, C., Wineland, D.: Quantum Dynamics of Single Trapped Ions. Rev. Mod. Phys. **75**(1), 281–324 (2003)
7. Blatt, R., Wineland, D.: Entangled States of Trapped Atomic Ions. Nature **453**(7198), 1008–1015 (2008)
8. Hanson, R., Kouwenhoven, L.P., Petta, J.R., Tarucha, S., Vandersypen, L.M.K.: Spins in Few-electron Quantum Dots. Rev. Mod. Phys. **79**(4), 1217–1265 (2007)
9. Awschalom, D., Loss, D., Samarth, N.: Semiconductor Spintronics and Quantum Computation. Springer, Heidelberg (2002)
10. Weber, J.R., Koehl, W.F., Varley, J.B., Janotti, A., Buckley, B.B., Van de Walle, C.G., Awschalom, D.D.: Quantum Computing with Defects. Proc. Natl. Acad. Sci. U.S.A. **107**(19), 8513–8518 (2010)
11. Prawer, S., Greentree, A.D.: Diamond for Quantum Computing. Science **320**(5883), 1601–1602 (2008)
12. Kane, B.E.: A Silicon-based Nuclear Spin Quantum Computer. Nature **393**(6681), 133–137 (1998)
13. Harrow, A.W., Hassidim, A., Lloyd, S.: Quantum Algorithm for Linear Systems of Equations. Phys. Rev. Lett. **103**(15), 150502 (2009)
14. Leyton, S.K., Osborne, T.J.: A Quantum Algorithm to Solve Nonlinear Differential Equations (2008). arXiv:0812.4423
15. Temme, K., Osborne, T.J., Vollbrecht, K.G., Poulin, D., Verstraete, F.: Quantum Metropolis Sampling. Nature **471**(7336), 87–90 (2011)
16. Somma, R.D., Boixo, S., Barnum, H., Knill, E.: Quantum Simulations of Classical Annealing Processes. Phys. Rev. Lett. **101**(13), 130504 (2008)
17. Givi, P.: Filtered Density Function for Subgrid Scale Modeling of Turbulent Combustion. AIAA J. **44**(1), 16–23 (2006)
18. Ansari, N., Jaberi, F.A., Sheikhi, M.R.H., Givi, P.: Filtered Density Function as a Modern CFD Tool. Eng. Appl. Comp. Fluid **1**(1), 1–22 (2011)
19. Yilmaz, S.L., Ansari, N., Pisciuneri, P.H., Nik, M.B., Otis, C.C., Givi, P.: Applied Filtered Density Function. J. Appl. Fluid Mech. **6**(3), 311–320 (2013)
20. Meier, W., Weigand, P., Duan, X.R., Giezendanner-Thoben, R.: Detailed Characterization of the Dynamics of Thermoacoustic Pulsations in a Lean Premixed Swirl Flame. Combust. Flame **150**(1–2), 2–26 (2007)
21. Ansari, N., Strakey, P.A., Goldin, G.M., Givi, P.: Filtered Density Function Simulation of a Realistic Swirled Combustor. Proc. Comb. Inst. **35**(2), 1433–1442 (2015)
22. Libby, P.A., Williams, F.A.: Turbulent Reacting Flows. Academic Press, London (1994)

23. Pope, S.B.: Turbulent Flows. Cambridge University Press, Cambridge (2000)
24. Givi, P.: Spectral and random vortex methods in turbulent reacting flows. In: Libby and Williams [22], chap. 8, pp. 475–572
25. Germano, M.: Turbulence: the Filtering Approach. J. Fluid Mech. **238**, 325–336 (1992)
26. Karlin, S., Taylor, H.M.: A Second Course in Stochastic Processes. Academic Press, New York (1981)
27. Gardiner, C.W.: Handbook of Stochastic Methods for Physics, Chemistry and the Natural Sciences, 2nd edn. Springer-Verlag, New York (1990)
28. Dopazo, C.: Recent Developments in PDF Methods. In Libby and Williams [22], chap. 7, pp. 375–474
29. Risken, H.: The Fokker-Planck Equation, Methods of Solution and Applications. Springer-Verlag, New York (1989)
30. Sheikhi, M.R.H., Givi, P., Pope, S.B.: Frequency-Velocity-Scalar Filtered Mass Density Function for Large Eddy Simulation of Turbulent Flows. Phys. Fluids **21**(7), 075102 (2009)
31. Martin, M.P., Piomelli, U., Candler, G.V.: Subgrid-Scale Models for Compressible Large-Eddy Simulations. Theor. Comp. Fluid Dyn. **13**(5), 361–376 (2000)
32. Gicquel, L.Y.M., Givi, P., Jaberi, F.A., Pope, S.B.: Velocity Filtered Density Function for Large Eddy Simulation of Turbulent Flows. Phys. Fluids **14**(3), 1196–1213 (2002)
33. Yilmaz, S.L., Nik, M.B., Sheikhi, M.R.H., Strakey, P.A., Givi, P.: An Irregularly Portioned Lagrangian Monte Carlo Method for Turbulent Flow Simulation. J. Sci. Comput. **47**(1), 109–125 (2011)
34. Pisciuneri, P.H., Yilmaz, S.L., Strakey, P.A., Givi, P.: An Irregularly Portioned FDF Simulator. SIAM J. Sci. Comput. **35**(4), C438–C452 (2013)
35. Sandia National Laboratories: TNF Workshop website, Bluff body flames (2015). www.sandia.gov/TNF/bluffbod.html
36. Ansari, N., Pisciuneri, P.H., Strakey, P.A., Givi, P.: Scalar-Filtered Mass-Density-Function Simulation of Swirling Reacting Flows on Unstructured Grids. AIAA J. **50**(11), 2476–2482 (2012)
37. Sammak, S., Ansari, N., Givi, P., Brazell, M.J., Mavriplis, D.J.: A DG-FDF Large Eddy Simulator. Bulletin of the American Physical Society **59**(20), 496 (2014)

Modeling of Adsorption and Transfer of Radiation in an Expanding Sphere

Nikolai Shaparev[1,2](✉)

[1] Institute of Computational Modeling, Russian Academy of Sciences,
Krasnoyarsk 660036, Russia
shaparev@icm.krasn.ru
[2] National Research Tomsk State University, Tomsk 634050, Russia

Abstract. In this paper we find spatial and average dependences of the optical medium thickness, spectral profile and absorption line width on the initial thickness of the medium and the ratio between the limiting velocity of self-similar expansion and the thermal velocity of atoms.

Keywords: Absorption spectrum · Self-similar gas expansion

1 Introduction

The studying of interaction between resonant radiation and a medium plays an important role in astrophysics and gas and plasma physics. Most of the research in this field has been devoted to stationary media. Sobolev [1] was the first to study the impact of macroscopic movement of gas on the transfer of resonance radiation in expanding planetary nebulae and to estimate the probability of photon escape from the medium. A further development of this approach can be found in [2]. The macroscopic movement is an important consideration when dealing with laser generation [3] or producing ultracold plasma (UCP) [4] in a laboratory.

Our previous research efforts were focused on the radiation transfer in expanding media [5–7], escape of resonance radiation from the center of an expanding medium [8], absorption of external continuum radiation in a sphere [9,10] and on finding an optimum laser radiation spectrum to obtain UCP [11].

In this paper we analyze the absorption spectrum characteristics (optical thickness, absorption line width and shape) in a self-similarly expanding gaseous sphere.

2 Model

In a self-similarly expanding gaseous [12] and plasma [4] sphere, the expansion velocity $\mathbf{V}(\mathbf{r})$ at the point $\mathbf{r}$ is given by

$$\mathbf{V}(\mathbf{r}) = \frac{V_R}{R}\mathbf{r}, \tag{1}$$

© Springer International Publishing Switzerland 2015
N. Danaev et al. (Eds.): CITech 2015, CCIS 549, pp. 133–142, 2015.
DOI: 10.1007/978-3-319-25058-8_14

where V_R is the gas expansion velocity at the boundary of the sphere of R radius. We will assume the gas concentration n and temperature T to be constant over space and the absorption line shape to be determined by the Doppler effect. Given these assumptions, the radiation absorption coefficient $k(\nu, \mathbf{r})$ at the frequency ν at the point $\mathbf{r}$ will have the form

$$k(\nu, \mathbf{r}) = k_0 \exp\left\{-\left[\frac{(\nu - \nu_0)}{\nu_0} - \frac{\mathbf{V}(\mathbf{r})\mathbf{l}}{c}\right]^2 \left(\frac{c}{V_0}\right)^2\right\}. \qquad (2)$$

Note that the first term in the exponent (2) is associated with thermal gas movement and results in a Doppler shape of the absorption line while the second term refers to macroscopic gas movement due to expansion. When there is no gas expansion, the absorption coefficient at the resonance frequency $\nu = \nu_0$ equals

$$k_0 = \frac{\lambda_0^3}{8\pi} \frac{g_2}{g_1} \frac{A_{21}}{\sqrt{\pi} V_0} n. \qquad (3)$$

Notations used in (2) and (3) are as follows: is the speed of light; $\lambda_0 = c/\nu_0$ is the resonance radiation wavelength; $\mathbf{l}$ is the unit vector determining the direction of propagation of the external radiation; $V_0 = \sqrt{2k_B T/m}$; k_B stands for the Boltzmann constant; m is the atom (ion) mass; g_2, g_1 are the statistic weights of the excited and ground state; A_{21} is the spontaneous decay probability.

Consider radiation absorption I_0 propagating along a chord that is parallel to the direction of travelling of radiation $\mathbf{l}$ and is determined by the angle φ formed from the center of the sphere between the direction $\mathbf{l}$ and the direction to the point where the chord intersects the surface of the sphere. For $\varphi = 0$, radiation propagates along the sphere diameter. By the Bouguer law, the intensity of radiation leaving a sphere at the frequency ν through unit area will be equal to $I(\nu, \varphi)$

$$I(\nu, \varphi) = I_0 \exp\left\{-k_0 \int_{R\cos\varphi}^{R\cos\varphi} \exp\left[-\left(\frac{(\nu - \nu_0)c}{\nu_0 V_0} - \frac{V_R}{V_0}\frac{r}{R}\right)^2\right] dr\right\}, \qquad (4)$$

where I_0 is the incident radiation intensity independent of frequency.

We now switch to new variables

$$\omega = \frac{(\nu - \nu_0)}{\nu_0}\frac{c}{V_0}, \qquad y = \frac{V_R}{V_0}\frac{r}{R}, \qquad (5)$$

and rewrite (4) as

$$I(\omega, \varphi) = I_0 \exp\left\{-\tau_1 \int_{-\alpha\cos\varphi}^{\alpha\cos\varphi} \exp\left[-(\omega - y)^2\right] dy\right\}, \qquad (6)$$

where

$$\tau_1 = \frac{\tau_0}{\alpha}, \qquad \tau_0 = k_0 R, \qquad \alpha = \frac{V_R}{V_0}. \qquad (7)$$

Here τ_0 is the optical thickness of the medium when there is no expansion and α is found as a ratio of the gas expansion velocity at the boundary to the thermal velocity of particles.

Finding a logarithm of (6) yields

$$\ln\left[\frac{I_0}{I(\omega,\varphi)}\right] = \tau_1 \int\limits_{-\alpha\cos\varphi}^{\alpha\cos\varphi} \exp\left[-(\omega-y)^2\right] dy. \tag{8}$$

After averaging over sphere, denoted as $<>$, expression (8) takes on the form

$$< \ln\left[\frac{I_0}{I(\omega,\varphi)}\right] >= 2\tau_1 \int\limits_{0}^{\pi/2} \int\limits_{-\alpha\cos\varphi}^{\alpha\cos\varphi} e^{-(\omega-y)^2} \sin\varphi\cos\varphi\, d\varphi. \tag{9}$$

Expressions (8) and (9) can provide spatial and mean estimates for the optical thickness of the medium and absorption line width and shape.

3 Results

3.1 Optical Thickness of a Medium

The optical thickness of an expanding medium is found from expression (8) for $\omega = 0$

$$\tau = \tau_1 \int\limits_{-\alpha\cos\varphi}^{\alpha\cos\varphi} e^{-y^2} dy \tag{10}$$

and depends on τ_0, φ and α. For radiation propagating along the diameter $(\varphi = 0)$, the optical thickness τ (10) is shown in Fig. 1 as a function of α. As we can see, the optical thickness reduces with growing α, which is attributed to the absorption spectrum broadening resulting from the expansion and to the associated reduced absorption at the frequency $\omega = 0$.

Consider asymptotic values of τ. For $\alpha\cos\varphi \ll 1$, after series expansion of the exponent in (10) followed by integration we obtain

$$\tau = 2\tau_0\cos\varphi\left[1 - \frac{\alpha^2\cos^2\varphi}{3}\right]. \tag{11}$$

For $\alpha\cos\varphi \gg 1$, from (10) we obtain

$$\tau = \sqrt{\pi}\frac{\tau_0}{\alpha}. \tag{12}$$

The asymptotic values τ shown in Fig. 1 provide fairly good estimates of τ for small and large α. According to (12), τ decreases when $\alpha > \sqrt{\pi}$ and the medium becomes transparent when $\alpha > \sqrt{\pi}\tau_0$.

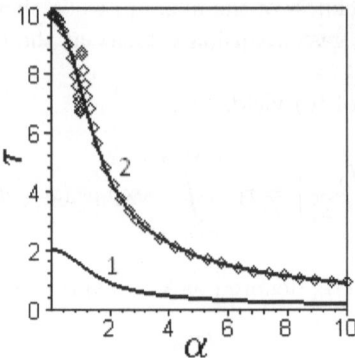

Fig. 1. Optical thickness for $\varphi = 0$. Curve 1 refers to $\tau_0 = 1$, Curve 2 to $\tau_0 = 5$. $\diamondsuit$ — asymptotic values: on the left as found from (11), on the right – from (12).

The simulations show that the optical thickness τ as a function of φ remains constant for $\alpha > \tau_0$ except for the near boundary areas ($\varphi \rightarrow \pi/2$) where $\tau = \tau_0/\alpha \rightarrow 0$ and hence the radiation is no longer absorbed. Explanation for the constant value of τ stems from τ_0 and α being proportionate to $\cos\varphi$.

According to (9) and (10), the optical thickness averaged over sphere will be

$$< \tau >= 2\tau_1 \int\limits_0^{\pi/2} \int\limits_{-\alpha\cos\varphi}^{\alpha\cos\varphi} e^{-y^2}\,dy \sin\varphi \cos\varphi\,d\varphi, \tag{13}$$

which is plotted in Fig. 2. Applying (11), for $\alpha \ll 1$ we obtain

$$< \tau >= 4\tau_0 \left[\frac{1}{3} - \frac{\alpha^2}{15}\right]. \tag{14}$$

For $\alpha \gg 1$, the averaging of (12) yields

$$< \tau >= \sqrt{\pi}\frac{\tau_0}{\alpha}. \tag{15}$$

Average asymptotic values $< \tau >$ are also shown in Fig. 2 illustrating good agreement between the asymptotic values and the numerical estimates. The average value of $< \tau >$ is considerably less than τ for $\varphi = 0$ because τ reduces along the chords as φ changes from 0 to $\pi/2$.

3.2   Absorption Line Shape

Frequency dependence (8) also describes the absorption line shape $I(\omega)$. For small α and $\varphi = 0$, series expansion of (8) with respect to y followed by integration yields

$$I(\omega) = 2\tau_0 e^{-\omega^2}\left[1 - \frac{1}{3}\left(2\omega^2 - 1\right)\alpha^2\right]. \tag{16}$$

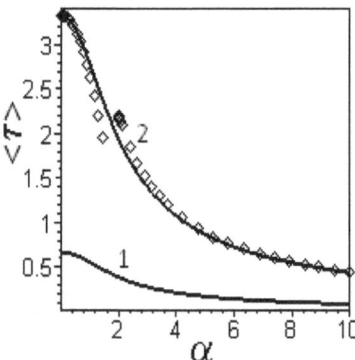

Fig. 2. Average optical thickness Curve 1 refers to $\tau_0 = 1$, Curve 2 to $\tau_0 = 5$. $\Diamond$ — asymptotic values: on the left as found from (14), on the right – from (15).

On the other hand, suppose V_0 increases by ΔV_0 , then, in virtue of the dependence of w and τ_0 on V_0, by expanding the Doppler line shape (Exp. (2) for $V_R = 0$) in series with respect to ΔV_0 we obtain

$$I(w) = 2\tau_0 e^{-w^2}\left[1 - \frac{(2w^2 - 1)}{V_0}\Delta V_0\right]. \tag{17}$$

Comparison of (16) and (17) indicates that $I(w)$ maintains Dopper distribution for small α and

$$\Delta V_0 = \frac{\alpha^2}{3}V_0, \tag{18}$$

which is associated with the gas expansion. The increase of V_0 by ΔV_0 correlates with the increase of the effective gas temperature by

$$\Delta T = \frac{2}{3}\alpha^2 T. \tag{19}$$

Fig. 3 shows $I(w)$ for $\alpha = 0.3$ and the Doppler distribution with a thermal velocity increased by ΔV_0. One can see that these distributions are quite close to each other.

Fig. 4 illustrates the absorption line shape corresponding to Exp. (8) when $\varphi = 0$ for $\tau_0 = 1$ and various α. It is obvious that the absorption line grows broader and acquires the shape of a plateau as the expansion velocity V_R and hence α increase. The profile gets narrower with increasing φ as the result of shrinking of the expansion velocity projection.

Note that the blue wing of the absorption curve ($w > 0$) is associated with atoms whose velocity projection V is parallel to l while the red wing ($w < 0$) is associated with atoms flying to meet radiation.

Formation of the plateau for large α is induced by the fact that the same number of particles is absorbed at each frequency. To prove this, let us put the

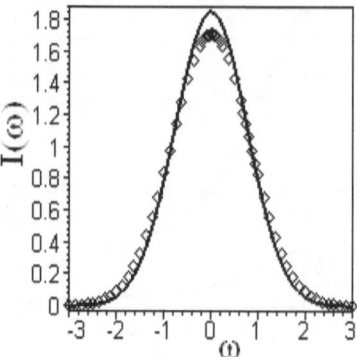

Fig. 3. Absorption line shape for $\varphi = 0$. The solid curve corresponds to Exp. (8) for $\alpha = 0.3$, $\diamond$ – Doppler distribution.

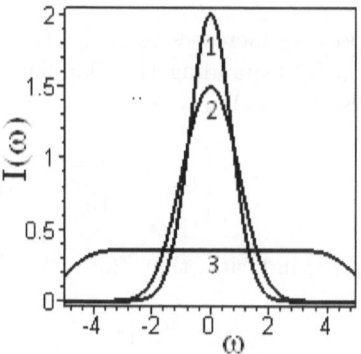

Fig. 4. Absorption line shape for $\varphi = 0$ and $\tau_0 = 1$. Curve 1: $\alpha = 0$, Curve 2: $\alpha = 1$, Curve 3: $\alpha = 5$.

ω derivative in (8) equal to zero for $\varphi = 0$

$$\int_{-\alpha}^{\alpha} e^{-(\omega-y)^2} 2(\omega - y) dy = 0. \tag{20}$$

By integrating (refeq:20) we obtain

$$e^{-(\omega-\alpha)^2} - e^{-(\omega+\alpha)} = 0. \tag{21}$$

Next we expand (21) in series with respect to ω in the range $\omega < \alpha$ and what we get is, indeed, equality to zero of expressions (20) and (21). The unaffected behaviour of $I(\omega)$ in the range $-\alpha < \omega < \alpha$ when $\alpha \gg 1$ is due to the fact that the medium gets optically thinner and the absorption of radiation is proportionate to n, the latter being constant in space.

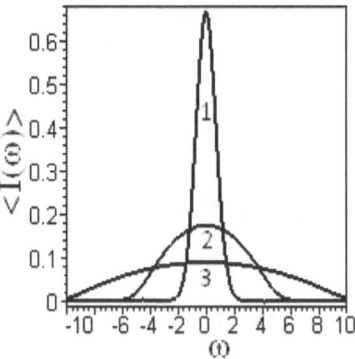

Fig. 5. Average shape of the absorption line. $\tau_0 = 1$; Curve 1 refers to $\alpha = 0$, Curve 2 to $\alpha = 1$, Curve 3 to $\alpha = 5$.

The plateau in the line shape disappears after averaging over, which brings in contributions from different projections of the expansion velocity. The average shape of the absorption line is plotted in Fig. 5.

3.3   Absorption Line Width

The absorption line width is determined by the frequency range equal to double frequency difference between $\omega = 0$ and ω_1 corresponding to half maximum absorption. For the Doppler line shape

$$\Delta\omega = 2\omega_1 = 2\sqrt{\ln 2}. \tag{22}$$

For small α, additional contribution to the velocity V_0 is found as (18). Then substituting ΔV_0 (18) into the expression for ω (5) and expanding in series with respect to α we obtain the line width

$$\Delta\omega = 2\sqrt{\ln 2}\left(1 + \frac{\alpha^2}{3}\right). \tag{23}$$

For $\alpha \gg 1$, expression (8) when $\omega = 0$ equals

$$I(0) \simeq \sqrt{\pi}\tau. \tag{24}$$

For

$$\omega \simeq \pm\alpha, \tag{25}$$

$$I(\pm\alpha) \simeq \frac{\sqrt{\pi}}{2}\tau. \tag{26}$$

Hence the line width in the given situation will be

$$\Delta\omega = 2\alpha. \tag{27}$$

140   N. Shaparev

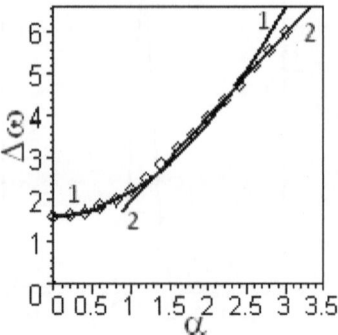

Fig. 6. Absorption line width for $\varphi = 0$. $\Diamond$ — as obtained from (8), Curve 1 as found from (23), Curve 2 – (27).

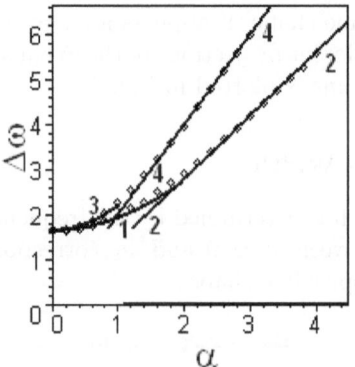

Fig. 7. Absorption line width (top – $\varphi = 0$, bottom – averaged over φ). Curve 1 refers to (29), Curve 2 to (31), Curve 3 – asymptotics (23), Curve 4 – asymptotics (27).

The line width as found from (8) is shown in Fig. 6 for $\varphi = 0$. Also shown are the asymptotic values employed: (23) and (27).

Let us now find the line width Δw averaged over sphere. For small α and $\varphi \neq 0$, the line width along the chord, by (23), will be

$$\Delta w = 2\sqrt{\ln 2} \left(1 + \frac{\alpha^2 \cos^2 \varphi}{3} \right) \tag{28}$$

By averaging (28) over φ we obtain

$$\Delta w = 2\sqrt{\ln 2} \left[1 + \frac{\alpha^2}{6} \right]. \tag{29}$$

For large α and $\varphi \neq 0$ we have

$$\Delta w = 2\alpha \cos \varphi. \tag{30}$$

After averaging we obtain

$$\Delta\omega = \frac{4}{3}\alpha. \tag{31}$$

Fig. 7 shows the line width for $\varphi = 0$ and the one averaged over φ. Also shown are the asymptotic values.

4   Conclusions

We have obtained spatial and average dependences of the optical thickness, shape and width of the resonance radiation absorption on the initial optical density τ_0 and the ratio of the limiting gas expansion velocity V_R to the thermal velocity of atoms V_0 ($\alpha = V_R/V_0$) in a self-similarly expanding gaseous sphere. The asymptotic and numerical estimates appear to be in good agreement for both small and large α.

The initial optical thickness has been shown to reduce with growing α until it becomes $-1/\alpha$ for large α due to frequency broadening of the absorption coefficient under expansion. This results in optical thinning when $\alpha > \tau_0$.

The Doppler shape of the absorption line is maintained for small α and the absorption line for $\alpha \gg 1$ in the frequency range $-\alpha < \omega < \alpha$ remains constant due to the dominating impact of expansion on the absorption spectrum. When averaged over sphere, the line shape becomes smooth. The absorption line broadening is associated with the Doppler shift due to macroscopic movement (expansion of the sphere).

The absorption line width along the chord is determined by $\Delta\omega \sim (1 + \alpha^2/3)$ for small $alpha$ and by $\Delta\omega \simeq 2\alpha$ for large α. Upon spatial averaging the coefficient in front of α decreases because of the changes in the velocity projection on the direction of radiation propagation.

Acknowledgments. The author is grateful to A.P. Gavriliuk for useful discussions of the results obtained.

References

1. Sobolev, V.V.: Moving Envelopes of Stars. Harvard University Press, Cambridge (1960)
2. Grinin, V.P.: Astrophysics **20**, 365 (1984)
3. ChenaisPopovies, C., Corbett, R., Hooker, C.J., Key, M.H., Kiehn, J.P., Lewis, C.L.S., Pert, J.P., Regan, C., Rose, S.J., Sadaat, S., Smith, A., Tomie, T., Willi, O.: Phys. Rev. Lett. **59**, 2161 (1987)
4. Killian, T.S., Pattard, T., Pohl, T., Rost, Y.M.: Phys. Reports **449**, 77 (2007)
5. Kosarev, N.I., Shaparev, N.Y.: J. Phys. B: Atom. Mol. and Opt. Phys. **44**, 316 (2011)
6. Kosarev, N.I., Shaparev, N.Y.: J. Phys. B: Atom. Mol. and Opt. Phys. **44**(5), 195402 (2011)
7. Kosarev, N.I., Shaparev, N.Y.: J. Phys. B: Atom. Mol. and Opt. Phys. **45**(5), 165003 (2012)
8. Shaparev, N.Y.: Doklady Physics **58**, 45 (2013)

9. Shaparev, N.Y.: J. Phys. B: Atom. Mol. and Opt. Phys. **47**(6), 22540 (2014)
10. Shaparev, N.Y.: Proceeding if International conference on optoelectronics and microelectronics, Changchun, China, vol. 337, p. 342 (2012)
11. Shaparev, N.Y.: Laser Phys. Lett. **10**(6), 985501 (2013)
12. Zeldovich, Y., Raizer, Y.: Physics of Shock Waves and High Temperature Hydrodynamic Phenomena. Academic Press, New York (1966)

Three-Dimensional Model of Fracture Propagation from the Cavity Caused by Quasi-Static Load or Viscous Fluid Pumping

Yuriy Shokin, Sergey Cherny, Denis Esipov, Vasily Lapin,
Alexey Lyutov, and Dmitriy Kuranakov[✉]

Institute of Computational Technologies of the SB RAS,
Acad. Lavrentjev, 6, 630090 Novosibirsk, Russia
{esipov,lapin,kuranakov}@ict.sbras.ru
http://www.ict.nsc.ru/ru/

Abstract. Fracture propagation caused by fluid pumping is in the focus of the report. The most popular approaches and problem statements used for the propagation simulation are described.

Methods of simulation of the main processes that take place during the fracture propagation are outlined. There processes are the follows: rock deformation and rock breaking, fluid flow inside the fracture and its filtration in the rock.

New method of fracture propagation simulation is proposed. The method unites three sub-models that describe three (except the fluid filtration) processes that affect the fracture propagation. Important advance of the methodic is its ability to replace any sub-model without numerical algorithm modification. So the appropriate sub-model can be chosen for each process depending on the problem features.

Thus quasi static and unsteady statement may be used for simulation of fracture propagation caused by viscous and inviscid fluid pumping. Rock deformation is described in scope of linear elasticity equation of homogeneous uniform material. Classical (similar to one used in [1]) and dual boundary element methods are used for this equations solution. Rock breaking caused by the fracture propagation is described by Irwin's criterion coupled with maximal circumferential stress criterion for calculation of propagation direction. Various approaches are used to obtain stress intensity factors that are necessary for both criteria.

Proposed methodic has been applied for fracture propagation simulation. The sensitivity of fracture propagation process to variation of the main physical parameters has been shown.

Keywords: Three-dimensional dual boundary elements method · Quasi-Static load · Viscous fluid · Hydraulic fracturing · Non-planar fracture propagation

1   Introduction

In the paper [2] a fully 3D numerical model of fracture propagation from the cavity in an elastic media caused by the viscous fluid pumping was developed

© Springer International Publishing Switzerland 2015
N. Danaev et al. (Eds.): CITech 2015, CCIS 549, pp. 143–157, 2015.
DOI: 10.1007/978-3-319-25058-8_15

and verified. Numerical model means linked submodels and numerical methods and algorithms for their coupled solution. Three basic submodels were linked together into a single model of propagation: the stress-strain state of the elastic media, Newtonian fluid flow, and brittle fracturing and crack growth. The following assumptions were made in the model [2]. The media fracturing velocity is assumed to be low enough. It allows using the fracture propagation the elastic equilibrium equations and the static criteria of crack growth and direction for the fracture propagation simulation. During the model of fracture propagation development authors of [2] were using conventional BEM for the stress-strain state calculations [3]. Therefore this method was used for the elasticity problem solution. However the conventional BEM cannot be used for the line cracks because the integral boundary equation degenerates. Therefore in paper [2] the fracture was considered as a cavity with small but finite width between its sides (Fig. 1). So the fracture was approximated by the crack with the artificial width, and the width itself was defined from the condition of the solution error minimization, caused by this fracture approximation.

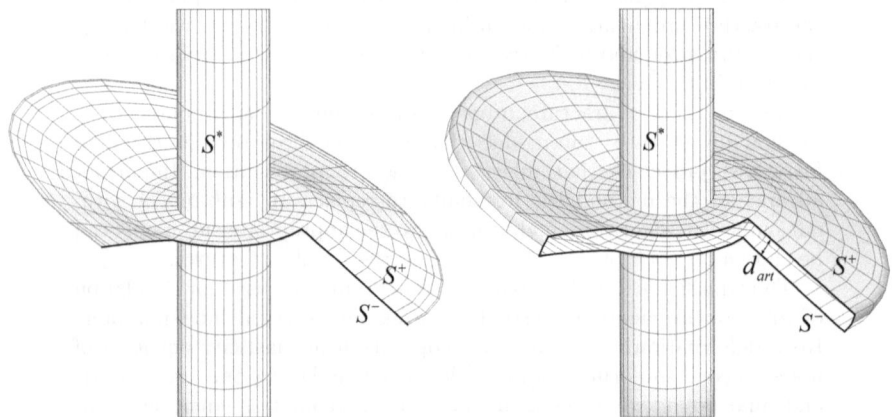

Fig. 1. Artificial notch concept: real fracture (left) is replaced with artificial notch (right) [2].

In the present paper the fracture is treated as a real crack with infinite small distance between sides. For the solution of the elasticity problem with the cavity and the fracture, the modification of the Dual BEM with discontinuous elements is built [4]. It is the most optimal method with regard to the computational costs and the convenience of the integral equations approximation. Near the crack front special elements are used. They account the singularity of the elasticity problem solution. To improve the accuracy of the Stress Intensity Factors calculation, the special boundary elements near the crack front are accounted in the interpolation formulae.

2   Dual BEM

The elasticity problem is solved in an infinite domain with the cavity bounded by S^* and the fracture with sides S^+ and S^- which adjoins the cavity (see Fig. 1, left). Stress-strain state of a media is described by elastic equilibrium equations

$$\frac{\partial \sigma_{ij}}{\partial x_j} = 0, \tag{1}$$

where σ_{ij} are the components of the stress tensor; indices i, j posses the values $1, 2, 3$. The Hookes law for the isotropic homogeneous material is used with the equation (1)

$$\sigma_{ij} = \lambda \delta_{ij} \varepsilon_{kk} + 2\mu \varepsilon_{ij}, \tag{2}$$

where $\varepsilon_{ij} = 0.5(u_{i,j} + u_{j,i})$ are the displacements tensor components, u_i are the components of the displacements vector, δ_{ij} -is the Kronecker symbol, λ and μ are the Lame parameters.

To obtain the closed differential problem let us add the boundary conditions on the cavity surface $S^* = S^t + S^u$

$$t_i \Big|_{S^t} = t_i^*, \quad u_i \Big|_{S^u} = u_i^*, \tag{3}$$

on the fracture sides $S^{\pm}$

$$t_i \Big|_{S^{\pm}} = -p_{crack} n_i, \tag{4}$$

and on the infinite distance

$$u_i \Big|_{S^{\infty}} = 0, \tag{5}$$

to the differential equations (1),(2).

Conventional BEM [5] is used to solve the elasticity problems with a regular boundary S^*. For the problems with fractures $S^{\pm}$ a modification of the conventional BEM – the Dual BEM is suggested in [4]. In DBEM the Displacements Boundary Integral Equation (DBIE) is solved on the regular boundary and the Traction Boundary Integral Equation (TBIE) is solved on the fracture boundary. To solve the elasticity problem near the fracture in an infinite elastic media a modification of DBEM is developed in the present paper.

For points $\mathbf{y}$ at the regular boundary S^* the DBIE is solved

$$c_{ij}(\mathbf{y}) u_i(\mathbf{y}) = \int_{S^*} U_{ij}(\mathbf{y}, \mathbf{x}) t_i(\mathbf{x}) dS(\mathbf{x}) -$$

$$-\fint_{S^*} T_{ij}(\mathbf{y}, \mathbf{x}) u_i(\mathbf{x}) dS(\mathbf{x}) - \int_{S^+} T_{ij}(\mathbf{y}, \mathbf{x}) \Delta u_i(\mathbf{x}) dS(\mathbf{x}). \tag{6}$$

The singular integrals $\fint$ and $\fint\!\!\!\!=$ are considered in the meaning of the Cauchy and Hadamard principal value, respectively. In DBEM on one side of the fracture the TBIE is taken instead of the DBIE.

$$t_j(\mathbf{y}^+) = \int_{S^*} L_{ij}(\mathbf{y}^+,\mathbf{x})t_i(\mathbf{x})dS(\mathbf{x})-$$

$$-\int_{S^*} M_{ij}(\mathbf{y}^+,\mathbf{x})u_i(\mathbf{x})dS(\mathbf{x}) - \fint_{S^+} M_{ij}(\mathbf{y}^+,\mathbf{x})\Delta u_i(\mathbf{x})dS(\mathbf{x}). \tag{7}$$

Here $L_{ij}(\mathbf{y}^+,\mathbf{x}) = D_{kij}(\mathbf{y}^+,\mathbf{x})n_k(\mathbf{y}^+)$ and $M_{ij}(\mathbf{y}^+,\mathbf{x}) = S_{kij}(\mathbf{y}^+,\mathbf{x})n_k(\mathbf{y}^+)$, and functions D_{kij} and S_{kij} are obtained from the kernels U_{ij} and T_{ij} by differentiation with respect to the corresponding coordinates and applying the Hookes law [6]. Equation (7) doesn't contain displacement components u_i on the fracture boundary, but allows to determine the unknown components of the displacement discontinuities Δu_i on the boundary.

2.1 Boundary Discretization and Obtaining the System of Linear Algebraic Equations (SLAE)

Let us demonstrate the numerical method of the TBIE (7) solving in the context of fracture $S = S^+ + S^-$. The whole fracture S is approximated with the boundary elements as it is shown in the Fig. 2

$$S \simeq \sum_{e=1}^{N_e} S^e. \tag{8}$$

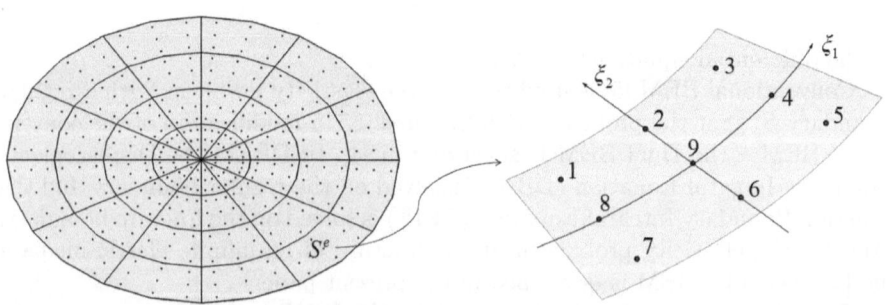

Fig. 2. Segmentation of boundary S into the discontinuous squared boundary S^e (left) and (ξ_1, ξ_2) parameterization of an element with $N_\alpha = 9$ (right).

Each boundary element S^e is parameterized with the local coordinates (ξ_1, ξ_2) as it is shown in Fig. 2. Components of the radius-vectors, displacements discontinuities, and stresses in a certain point of an element (ξ_1, ξ_2) are represented as

$$f_i(\xi_1, \xi_2) = \sum_{\alpha=1}^{N_\alpha} f_i(\mathbf{x}^\alpha)\phi_\alpha(\xi_1, \xi_2), \tag{9}$$

where $\mathbf{x}^\alpha$ are the element nodes, $\phi_\alpha(\xi_1, \xi_2)$ are the element shape functions, N_α is the number of nodes and shape functions in the element.

Equation (7) with respect to the (8) and (9) can be written as

$$t_j(\mathbf{y}^-) = \sum_{e=1}^{N_e} \sum_{\alpha=1}^{N_\alpha(e)} \left(-\Delta u_i^{e\alpha} \int_{\xi_1} \int_{\xi_2} M_{ij}(\mathbf{y}^-, \xi_1, \xi_2) \phi_\alpha(\xi_1, \xi_2) J(\xi_1, \xi_2) d\xi_1 d\xi_2 \right),$$

(10)

where $J(\xi_1, \xi_2)$ is the Jacobian of the transition to the elements local coordinate system. The displacements discontinuities $\Delta u_i^{e\alpha}$ in the node α of the element e are taken outside the integral because they doesnt depend on the integration variables ξ_1 and ξ_2. Note that integrals in formula (10) depend only on boundary geometry and not on the boundary conditions.

By writing out the equations (10) in the nodes $\mathbf{y}^{e\alpha}$, SLAE for the unknown functions Δu_i is obtained

$$\mathbf{M}\Delta\mathbf{u} = -\mathbf{t}.$$

(11)

Here the $\Delta\mathbf{u}$ and $\mathbf{t}$ are the vectors of the displacement discontinuities and tensions in all of the nodes. $\mathbf{M}$ is the matrix, composed of the integral values in equation (10).

In case with cavity and fracture $S = S^* + S^+ + S^-$ the system (11) is written as

$$\begin{bmatrix} \mathbf{T}_{11} - \frac{1}{2}\mathbf{I} & \mathbf{T}_{12} \\ \mathbf{M}_{21} & \mathbf{M}_{22} \end{bmatrix} \begin{pmatrix} \mathbf{u} \\ \Delta\mathbf{u} \end{pmatrix} = \begin{bmatrix} \mathbf{U}_{11} & 0 \\ \mathbf{L}_{21} & -\mathbf{I} \end{bmatrix} \begin{pmatrix} \mathbf{t} \\ \mathbf{t} \end{pmatrix},$$

(12)

where $\mathbf{U}$ and $\mathbf{T}$ are the sub-matrices, composed of the integral values in the DBIE (6), $\mathbf{L}$ and $\mathbf{M}$ are the sub-matrices of the TBIE (7).

2.2   Boundary Elements and Approximating Functions

As long as TBIE (7) requires the smoothness of the surface in the collocation points $\mathbf{y}$ on the fracture $S^\pm$, and the elements edges are the lines of discontinuity, DBEM requires to use discontinuous elements with all nodes situated inside the element as it is shown in Fig. 2. In the present paper the discontinuous linear and squared elements, and special elements for the fracture front were used [7]. These elements approximate the displacement discontinuity $\Delta\mathbf{u}$ asymptotic at the fracture front, which improves the accuracy of the Stress Intensity Factors calculations.

2.3   Hadamar Principal Value Calculation of the Singular Integral

The main difficulty of DBEM is to construct the algorithm for the calculation of the Hadamar principal value for the singular integral along the boundary element S^e that appear in the equation (8). The integral contains the collocation point $\mathbf{y}$

$$I_{ij}(\mathbf{y}) = \oiint_{S^e} K_{ij}(\mathbf{y}, \mathbf{x}) dS(\mathbf{x}).$$

(13)

To calculate the integral I_{ij} (13) the singularity substraction technique [8] is used.

3   Calculation of the Stress Intensity Factors

The fundamental postulate of Linear Elastic Fracture Mechanics (LEFM) is that the behaviour of cracks is determined solely by the value of the Stress Intensity Factors (SIFs). The stress field in the vicinity of the crack tip is characterized by the SIFs K_I, K_{II} and K_{III}. In the present paper the displacement extrapolation method for evaluating SIFs is employed [6]

$$K_I^O = \frac{E}{4(1-\nu^2)}\sqrt{\frac{\pi}{2l}}\Delta u_b^P, \tag{14}$$

$$K_{II}^O = \frac{E}{4(1-\nu^2)}\sqrt{\frac{\pi}{2l}}\Delta u_n^P, \tag{15}$$

$$K_{III}^O = \frac{E}{4(1+\nu)}\sqrt{\frac{\pi}{2l}}\Delta u_t^P, \tag{16}$$

where $\Delta \mathbf{u}^P$ is the displacement discontinuity in the fracture point P placed at the distance l from a front point O. Vectors $\mathbf{b}$, $\mathbf{n}$ and $\mathbf{t}$ are local basis on the crack front. Formulae (14), (15), (16) are applicable if the distance l is small enough comparing to the typical fracture size. If the distance l is long, then the SIFs values become understated. In this case the extrapolation of the SIFs values K^{P_1} and K^{P_2}, from the points P_1 and P_2 to the front point O should be used (Fig. 3). Distance to the P_1 and P_2 is l_1 and l_2, respectively.

$$K^O = K^{P_2} + \frac{l_2(K^{P_1} - K^{P_2})}{l_2 - l_1}. \tag{17}$$

To verify the DBEM and the SIFs calculation method the following problem is considered. In the infinite media stretched by tensile stress σ in the direction y the penny-shaped fracture of radius R is placed. The fracture is inclined around the Oz axis at the angle α as it is shown in Fig. 4. The SIFs on the crack front for this problem were previously determined exactly [9]

$$K_I = 2\sigma \cos^2 \alpha \sqrt{\frac{R}{\pi}}, \tag{18}$$

$$K_{II} = \frac{4}{2-\nu}\sigma \sin \alpha \cos \alpha \cos \theta \sqrt{\frac{R}{\pi}}, \tag{19}$$

$$K_{III} = \frac{4(1-\nu)}{2-\nu}\sigma \sin \alpha \cos \alpha \sin \theta \sqrt{\frac{R}{\pi}}, \tag{20}$$

where θ is the angular coordinate characterizing the position of the point at the fracture front.

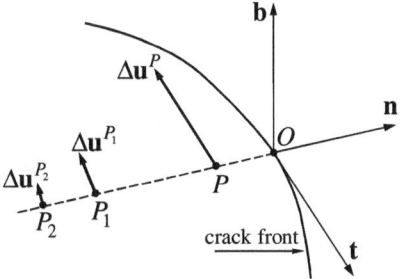

Fig. 3. Method of the displacements extrapolation near the crack front for the SIFs calculation.

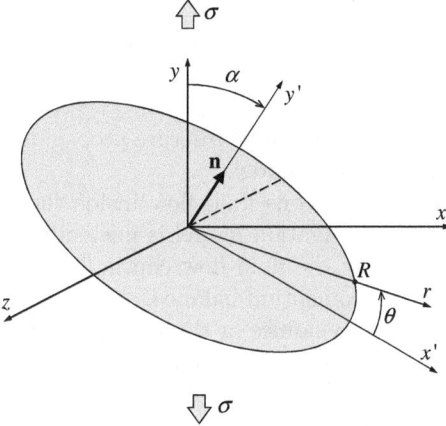

Fig. 4. Problem of a penny-shaped fracture inclined by an angle α under tensile stress σ

Problem is solved numerically on the mesh with 64 elements in the circumferential direction and 16 elements in radial direction. The physical parameters values are $R - 1m$, $p - 1MPa$, $E - 20GPa$, $\nu - 0.2$, $\alpha - 45°$. Figure 5 shows the distribution of the SIFs along the crack front. SIFs are calculated using the special elements and formula (17). Computational error does not exceed 2%.

4 Quasi-Static and Viscous Fluid Fracture Loading

In the case of the high confining stress of deep reservoirs and the low fluid viscosity the fluid pressure along fracture faces is nearly constant. Therefore two models of fracture loading are considered.

In the first one we assume that the fluid pressure is constant along the fracture faces, although it can be time-dependent. Under this condition it is also assumed that fluid and fracture fronts coincide, i.e., the size of so-called fluid lag

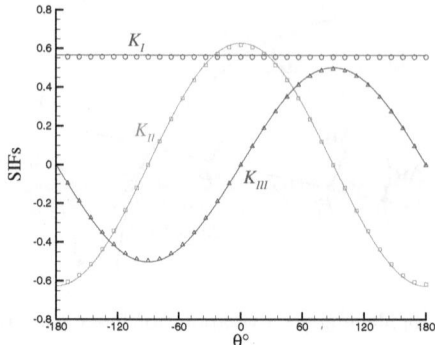

Fig. 5. Dependence of the SIFs from the point position at the front of a penny-shaped fracture: exact solution (solid); K_I (◯), K_{II} (□), K_{III} (△).

is negligible. We consider that hydraulic fracture propagation regime is described by the quasi-static crack growth model.

In the second model the viscous fluid flow inside the fracture is taken into account. In this case the propagation model is unsteady. The process unsteadiness is taken into account by the fluid-flow continuity equation. Meanwhile all other equations describing momentum balance, elastic equilibrium, and material rapture are stationary. The dynamics of the propagation process is represented by the static conditions of flow momentum, stress field, and elastic media displacements in various moments of time.

Fracture surface in 3D space and its piecewise planar representation is shown in Fig. 6. Through the boundary S^q fracturing fluid is pumped from the wellbore to the crack. Boundary S^p is the fluid's front.

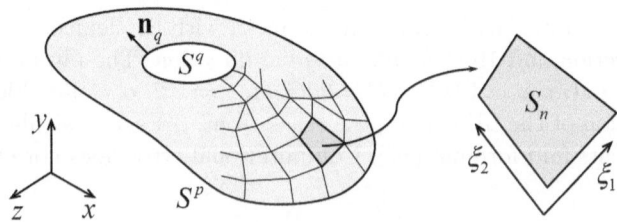

Fig. 6. Fracture surface in 3D space and its piecewise planar representation.

At each planar fracture element the lubrication approximation for a Newtonian fluid flow of viscosity μ between parallel plates, with distance W between each other, gives

$$\mathbf{q} = -\frac{W^3}{12\mu}\nabla p \tag{21}$$

where $\mathbf{q}$ is fluid flux.

The mass conservation equation can be written as follows

$$\frac{\partial W}{\partial t} + \nabla \cdot \mathbf{q} = 0. \tag{22}$$

From (21) - (22) it is possible to obtain the following equation for p:

$$\nabla(a\nabla p) = f, \tag{23}$$

where $a = \frac{W^3}{12\mu}$, $f = \frac{\partial W}{\partial t}$.

Boundary conditions for the equation (23) are the following:

$$p\Big|_{S^p} = p_{pore} \tag{24}$$

and the inflow condition is

$$\int_{S^q} \mathbf{q} \cdot \mathbf{n}_q dS = Q_{in}, \tag{25}$$

Here $\mathbf{n}_q$ is the normal to the boundary S^q. In terms of the pressure the latter condition (25) with consideration of (21) is rewritten as

$$\int_{S^q} a\frac{\partial p}{\partial n} dS = -Q_{in}. \tag{26}$$

It is considered that the fluid front moves with the same speed $\mathbf{v}_f$, as the fluid particles $\mathbf{v}(\mathbf{x})$ at the front do (Stefan condition)

$$\mathbf{v}_f(\mathbf{x}) = \mathbf{v}(\mathbf{x}) = \mathbf{q}(\mathbf{x})/W(\mathbf{x}), \quad \mathbf{x} \in S^p. \tag{27}$$

5 Coupling Between Stress-Displacement, Fluid-Flow and Crack Growth Criteria

Let us consider an initial fracture with front defined by the points $\mathbf{x}_i^0$, $i = 1, ..., N_{fr}$. Step-by-step fracture propagation is denoted by superscript n. Fluid front with nodes $\mathbf{x}_{f\ i}^n$, fracture front $\mathbf{x}_{r\ i}^n$, and the lag $L_{r\ i}$ between the fluid and the fracture fronts are introduced into the propagation algorithm. Also the volume V^n of the fluid in the fracture is interacting in the algorithm. It is calculated using the fracture width as

$$V^n = \int_{S+} W^n dS. \tag{28}$$

The general scheme of the propagation algorithm is shown in Fig. 7. The hydrodynamics-elasticity problem in the algorithm gives the distribution of the fracture width $W^{n+1\ s}$ and the pressure $p^{n+1\ s}$. Pressure is caused by the fluid flow in the fracture at the fracture front position $\mathbf{x}_{r\ i}^{n+1\ s}$ and the fluid front position $\mathbf{x}_{f\ i}^{n}$. The scheme of the hydrodynamics-elasticity problem solution is shown in Fig. 8. Iteration process $\Delta t^{k+1} = \mathbb{T}(\Delta t^k)$ is introduced to fulfill the condition

$$\max_i \left| \mathbf{v}_i^{m+1\ k} \right| = v_f, \tag{29}$$

which equalizes the maximal fluid velocity at the front and the kinematic condition of the given maximal front increment L_f^0 over the time period Δt that is calculated from the fracture volume dynamics.

With the iterations

$$L_i^{s+1} = \mathbb{L}(L_i^s), \quad \theta_i^{s+1} = \mathbb{Q}(\theta_i^s) \tag{30}$$

the following conditions are fulfilled in the algorithm Fig. 7

$$K_I(\mathbf{x}_i^{n+1\ s}, p^m) = K_{Ic}, \quad K_{II}(\mathbf{x}_i^{n+1\ s}, p^m) = 0 \tag{31}$$

in each of the fracture front nodes on the $n + 1$-th propagation step. Iteration schemes (30) are based on the solution methods for the equations (31) correspondingly.

6   Results of Fracture Propagation Simulating

Figures 9 – 11 show the simulation results of the quasi-static propagation of the penny-shaped fracture with radius R from the wellbore with radius R_w. The initial fracture is perpendicular to the axis of the wellbore, which is inclined at the angle α to the vertical direction (axis Oy). Parameter values during the simulation are $E = 20GPa$, $\nu = 0.2$, $K_{Ic} = 3MPa\sqrt{m}$, $R = 1m$, $R_w = 0.5m$, $\alpha = 30°$.

The isometric projections of the fracture during the quasi-static propagation are shown in Fig. 11. The fracture is propagating with constant in situ stress $\sigma_x^\infty = \sigma_z^\infty = 16MPa$, and various in situ stress $\sigma_y^\infty = 8MPa$ (left) and $15.9MPa$ (right). The trajectories in the plane $z = 0$ are also compared in the figure.

The comparison of the quasi-static and the fluid-flow approach to the simulation of the fracture propagation is shown in Fig. 12. Wellbore is inclined against the σ_y^∞ direction at the angle $\alpha = 45°$ as it is shown in Fig. 9. Fluid with viscosity μ is pumped into the wellbore with rate $Q_{in} = 1 \cdot 10^{-3} m^3/s$. Rock is compressed by vertical $\sigma_y^\infty = 12MPa$ and two horizontal $\sigma_x^\infty = 16MPa$ and $\sigma_z^\infty = 16MPa$ stresses. The wellbore height and radius are $H = 5m$, $R_w = 0.5m$. The incipient fracture radius is $R = 1m$. The dynamic fluid flow approach is applied with the two values of fluid viscosity $\mu = 100$ and $1000Pa \cdot s$.

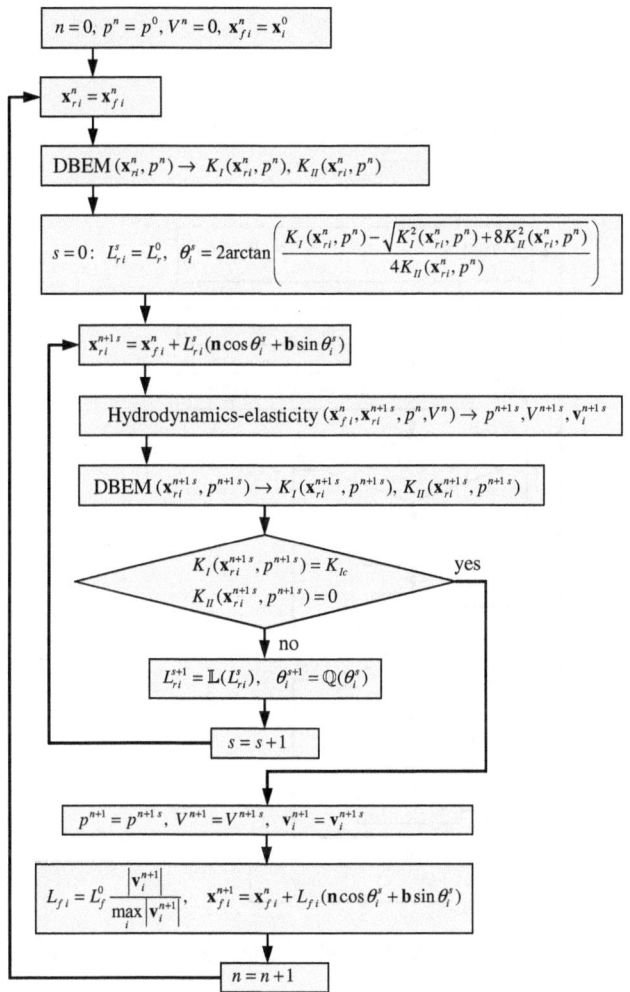

Fig. 7. The fracture propagation algorithm flow chart.

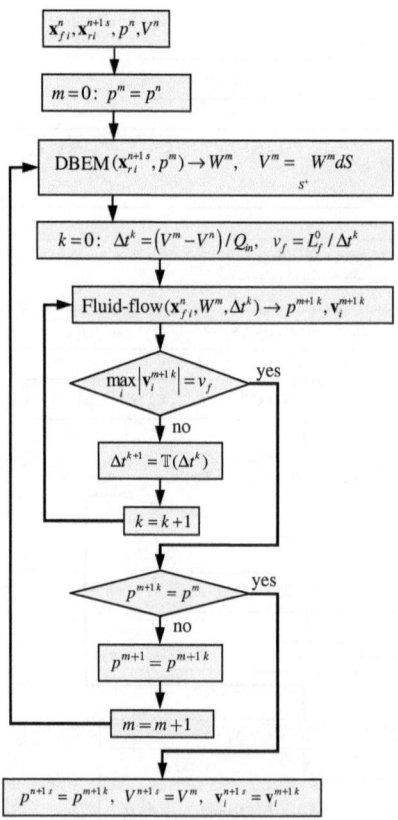

Fig. 8. The flowchart for the hydrodynamics-elasticity problem solution.

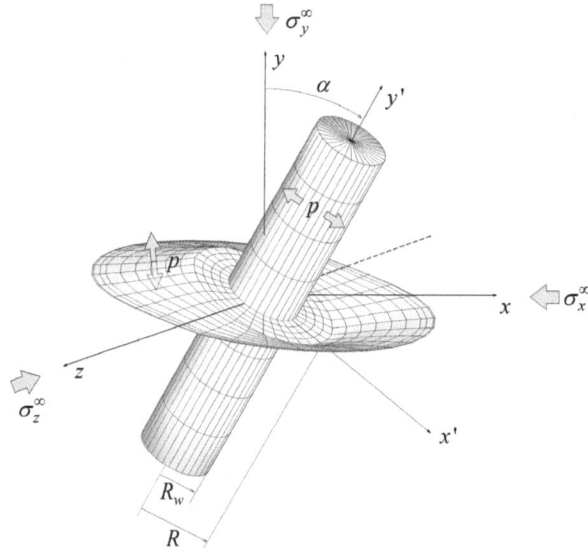

Fig. 9. Cavity and fracture loaded with pressure p in a media, which is compressed by a tensor $\boldsymbol{\sigma}^\infty$ on an infinite distance: $\sigma_x^\infty = -16MPa$, $\sigma_y^\infty = -12MPa$; $\sigma_z^\infty = -16MPa$

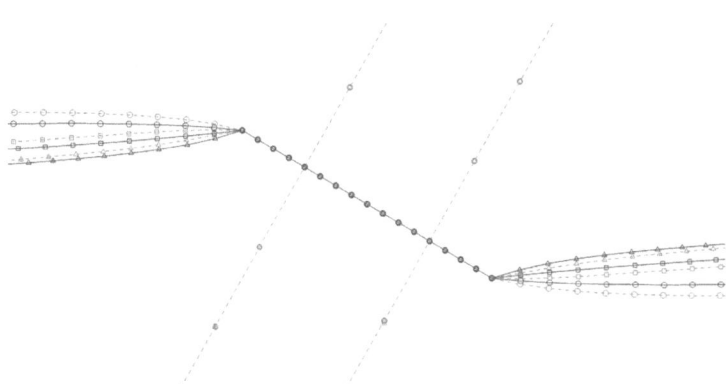

Fig. 10. Fracture trajectories in problems with the wellbore (dashed line) and without (solid line): $(\sigma_x^\infty; \sigma_y^\infty; \sigma_z^\infty) = -(4; 3; 4)MPa$ ($\bigcirc$), $-(8; 6; 8)MPa$ ($\square$), $-(16; 12; 16)MPa$ ($\triangle$).

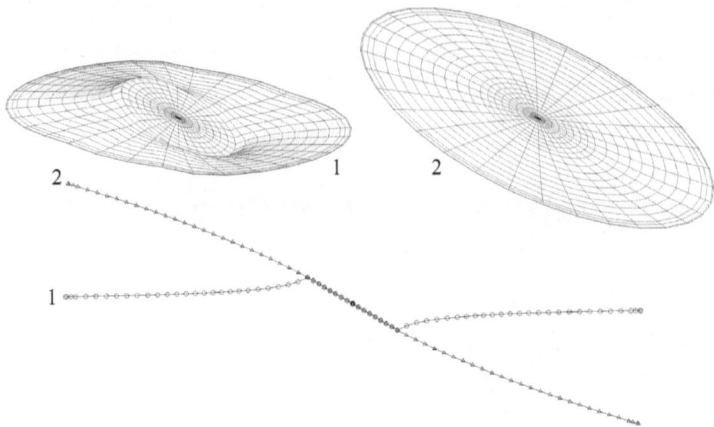

Fig. 11. The quasi-static fracture propagation: $1 - \sigma_y^\infty = 8MPa$ (left); $2 - \sigma_y^\infty = 15.9MPa$ (right); the trajectories in the section $z = 0$ (bottom).

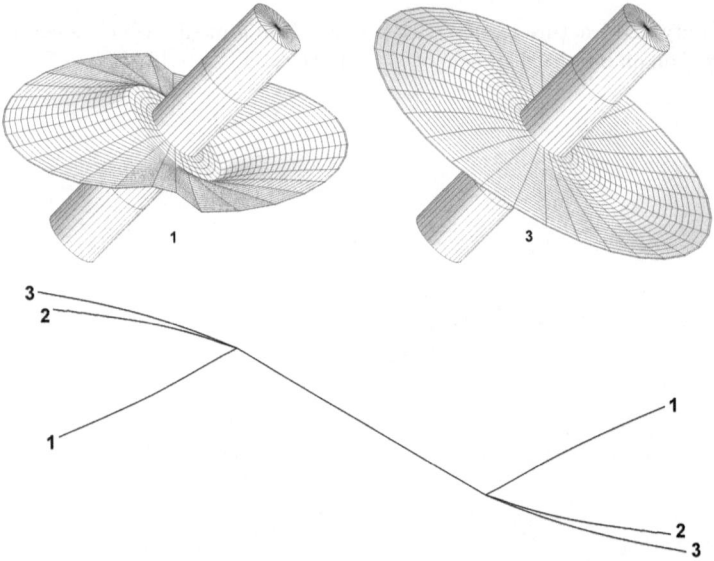

Fig. 12. Fracture trajectories and their cross-sections: 1 - quasistatic approach; 2 - dynamic approach $\mu = 100Pa \cdot s$; 3 - dynamic approach $\mu = 1000Pa \cdot s$

7   Conclusions

1. The concept of the 3D non-planar model of fracture propagation in an elastic media and the numerical algorithm for its implementation are proposed.

2. The concept combines models of the main linked problems that affect one another: stress-strain state, fracture loading, destruction of material, and fracture propagation.

3. The main advantage of the proposed conception is the possibility of using various models in every sub-problem without the necessity to rebuild the whole algorithm, which allows advancing from simple models to complex ones easily.

4. The version of the model that combines the sub-models of the elastic equilibrium, Newtonian fluid flow, and the fracture propagation and direction criterion derived from the linear brittle fracture mechanics is implemented.

5. The verification of the model and the sensitivity analysis of the solution from physical and numerical parameters is performed. It is shown that the results obtained are reliable.

6. The next version of the model will use more precise algorithms of SIFs calculations; the Newtonian fluid will be replaced with the non-Newtonian.

Acknowledgments. Authors gratefully acknowledge the support of this research by the Russian Scientific Fund under grant number 14-11-00234.

References

1. Lapin, V.N., Cherny, S.G., Esipov, D.V., Kuranakov, D.V.: 3D model of fracture initiation and propagation from the cavity in the elastic media loaded by constant pressure. In: Proceedeings of VIII Kazachstan-Russian Conference "Mathematical Modelling in Science and Technical Problems of Oil and Gas Industry", vol. 2, pp. 129–132, Jun 20–21, Kazakhstan, Atyrau (2014) (in Russian)

2. Cherny, S.G., Lapin, V.N., Esipov, D.V., Kuranakov, D.S., Avdyushenko, A.Y.: Simulating fully 3D non-planar evolution of hydraulic fractures. Submitted to the International Journal of Fracture (2015)

3. Alekseenko, O.P., Potapenko, D.I., Cherny, S.G., Esipov, D.V., Kuranakov, D.S., Lapin, V.N.: 3D Modeling of fracture initiation from perforated non-cemented wellbore. SPE J. **18**(3), 589–600 (2013)

4. Mi, Y., Aliabadi, M.H.: Dual boundary element method for three-dimensional fracture mechanics analysis. Engineering Analysis **10**(2), 161–171 (1992)

5. Rizzo, F.J.: An Integral Equation Approach to Boundary Value Problems of Classical Elastostatics // Quart. J. of Applied Mathematics **25**, 83–95 (1967)

6. Aliabadi, M.H.: The Boundary Element Method. Applications in Solids and Structures, vol. 2, 598p. John Wiley and Sons Ltd. (2002)

7. Cisilino, A.P., Aliabadi, M.H.: Three-dimensional BEM analysis for fatigue crack growth in welded components. Int. J. for Pressure Vessel and Piping **70**, 135–144 (1997)

8. Guiggiani, M., Krishnasamy, G., Rudolphi, T.J., Rizzo, F.J.: A general algorithm for numerical solution of hypersingular equations. J. Appl. Mech. **57**, 906–915 (1990)

9. Tada, H., Paris, P., Irwin, G.: The Stress Analysis of Cracks Handbook, 3rd edn. ASME Press, NY (2000)

Self-Purification Modelling for Small River in Climate Conditions of Central Siberia

Olga Taseiko[1](✉), Tatyana Spitsina[2], and Hranislav Milosevic[3]

[1] Siberian State Aerospace University, Krasnoyarsky Rabochy Av., 31, Krasnoyarsk,
Russia
taseiko@gmail.com
[2] Siberian State Technological University, Mira Av., 82, Krasnoyarsk, Russia
[3] Faculty of Science and Mathematics, University of Pristina,
Lole Ribara b.b., 38220 Kosovska Mitrovica, Serbia

Abstract. Water quality modelling in small rivers is often considered
unworthy from a practical and economic point of view. This work shows
that a simple model structure can be set up to describe the stationary water quality in small river basins in terms of carbon and nitrogen compounds, when it is unfeasible to use complex models. The
one-dimensional model include principle factors such as chemical and
biological oxidation, concentration of nutrients. Natural process of self-
purification for small river in sharp continental climate of Central Siberia
is inhibited by low temperatures, rapid currents and poor development
of plankton cenosis. So, a determination of model parameters demands
carrying out of special experiments with water samples. The results of
numerical modelling are verified by data from the environmental monitoring of some rivers in the basin of Central Enisey.

Keywords: Self-purification modelling · Reaeration rate · Biochemical
degradation processes

1 Introduction

Water quality management is usually affected by a variety of uncertainties raising
from the hydrodynamic conditions and meteorological processes, the variability
in the pollutant transport, the physicochemical processes, the indeterminacy of
available water and treated wastewater. While the number of models is stag-
gering, the fundamental concepts on which they are based are similar. Water
quality models represent the following: the hydrodynamic flow fields that drive
the movement of the water quality constituents, the movement and transformations of the water quality constituents [1] [2]. Eutrophication plays important
role in these models. Eutrophication is a process in which a water becomes rich
of nutrients (nitrogen, phosphorus, etc.), from domestic drainage as well as water
from agricultural practices. Production of oxygen is decreased in the water body
due to these processes.

© Springer International Publishing Switzerland 2015
N. Danaev et al. (Eds.): CITech 2015, CCIS 549, pp. 158–165, 2015.
DOI: 10.1007/978-3-319-25058-8_16

In Europe water quality models are far less prevalent in the regulatory process. Therefore, modelling the quantity of flow in the river is generally more important than modelling the quality. Nonetheless, there is a gradually increasing emphasis on quality modelling. UK environmental agencies use simple stochastic models to help the agencies decide on future restoration activities or permit for dischargers. Monte Carlo simulation is incorporated in the procedure to compensate for the inherently large uncertainty in the sparse data set [3].

In Russia, the flow and transport models are well developed and commonly used for engineering purpose. These models calculate hydrology and hydrodynamic conditions (flow, velocity, surface runoff), movement of water quality constituents. The problems of eutrophication are well studied for lakes and impoundments [4] [5] [6] [7] [8].

In developing water quality models small river basins pose specific problems due to data scarcity and the large number of diverse inputs, especially if they flow through urbanized territory. In these cases, it makes sense to use simple models in order to derive the crucial information about the river quality.

It is known that the simple models are easier to calibrate and therefore more reliable, but complex models are generally very sensitive and therefore difficult to identify all parameters [7]. Moreover, large rivers are more likely to be dominated by transport and conversion processes.

The most parameters of all models depend on regional specialty: the river characteristics (water flow, river bed morphology, depth), climate (temperature of air and water, precipitation), percentage of forest land, availability of groundwater etc [9].

The purpose of this work is to develop mathematical model for small rivers eutrophication taking into account the regional conditions.

2   Object Description

The Kacha river is considered in this study. It is the river in the basin of Central Enisey. Hydraulically the river is subjected to spring flood, but water level reduces significantly during the summer months, when the river quality becomes critical and the self-purification processes are almost stopped.

Sharp continental climate of Central Siberia, basin geology and vegetation define hydrological conditions of river flow. So, river flow rate and flow velocity differ significantly in various hydrological stages. All factors define regional features of eutrophication processes.

For model verification we use the data from state monitoring network for the period since 1985 to 2010 in Kacha river (three river station). All parameters are sampled in the basic hydrological stages (7-9 times in year). In this work we use the concentrations of oxygen, nitrogen, phosphorus and their compounds. Additionally, some complex parameters were measured in Kacha river during 2013 - 2015: pH, dissolved oxygen, biochemical oxygen demand (BOD), redox potential and conductivity. These parameters measured two times in week during period without ice cover.

3   Model Structure

The general one-dimensional advective-diffusive dynamics for a reactive pollutant can be written as a differential equations [10] [11] [12]. This one-dimensional in the x-direction model can be appropriate only for small rivers that is characterizing by small fluctuations on vertical and horizontal coordinates. This assumption wouldn't be appropriate for large rivers. Two- and three-dimensional representations are also possible, but they have considerable computational complexity. Neglecting the diffusion term yields [13]:

$$\frac{d(\omega \cdot C_i)}{dt} + \frac{d(Q \cdot C_i)}{dx} = K_{C_i} \cdot C_i \tag{1}$$

where K_{C_i} is decay rate of pollutant, that characterizes transformation velocity defined by the influence of chemical and biological processes, Q is river flow rate, ω is cross-sectional area of river.

The transformational processes included in the model are: degradation of dissolved carbon substances, ammonium oxidation, phosphorus mineralization, denitrification, and dissolved oxygen balance, including depletion by degradation processes and supply by physical reaeration and biochemical oxidation production.

The model includes equations for concentration some parameters: phosphate C_{PO_4}, total phosphorus C_{DOP}, ammonium nitrogen C_{NH_4}, nitrate nitrogen (including nitrite nitrogen) C_{NO_3}, total nitrogen C_{DON}, biochemical oxygen demand C_{org}, dissolved oxygen C_{O_2}. BOD characterizes oxygen's equivalent for dissolved organic carbonaceous demand.

Moreover, the modelling system incorporates transformation rates of all substances $K_i(day^{-1})$ and overland surface runoff $G_i(g/(m \cdot day))$. Surface runoff results the transport of pollutants into receiving waters via overland surface runoff within a drainage basin.

System equation are based on the equation (1):

1) Phosphate (gP/m^3):

$$\frac{d(\omega \cdot C_{PO_4})}{dt} + \frac{d(Q \cdot C_{PO_4})}{dx} = G_{PO_4} + K_{PO_4} \cdot \omega \cdot C_{DOP} \tag{2}$$

where Q is river flow rate (m^3/day), ω is cross-sectional area of river (m^2), K_{PO_4} is mineralization rate of total phosphorus (day^{-1}).

2) Total phosphorus (gP/m^3):

$$\frac{d(\omega \cdot C_{DOP})}{dt} + \frac{d(Q \cdot C_{DOP})}{dx} = G_{DOP} - K_{PO_4} \cdot \omega \cdot C_{DOP} \tag{3}$$

3) Ammonium nitrogen (gN/m^3):

$$\frac{d(\omega \cdot C_{NH_4})}{dt} + \frac{d(Q \cdot C_{NH_4})}{dx} = G_{NH_4} + K_{NH_4} \cdot \omega \cdot C_{DON} - K_{12} \cdot \omega \cdot C_{NH_4} \tag{4}$$

where K_{NH_4} is nitrogen mineralization rate (day^{-1}), K_{12} is nitrification rate (day^{-1}).

4) Nitrate nitrogen (gN/m^3):

$$\frac{d(\omega \cdot C_{NO_3})}{dt} + \frac{d(Q \cdot C_{NO_3})}{dx} = G_{NO_3} + K_{12} \cdot \omega \cdot C_{NH_4} - K_{NO_3} \cdot \omega \cdot C_{NO_3} \quad (5)$$

where K_{NO_3} is denitrification rate (day^{-1}).
   5) Total nitrogen (gN/m^3):

$$\frac{d(\omega \cdot C_{DON})}{dt} + \frac{d(Q \cdot C_{DON})}{dx} = G_{DON} - K_{NH_4} \cdot \omega \cdot C_{DON} \quad (6)$$

   6) BOD (gO_2/m^3):

$$\frac{d(\omega \cdot C_{org})}{dt} + \frac{d(Q \cdot C_{org})}{dx} = -K_{BOD} \cdot \omega \cdot C_{org} - K_{NO_3} \cdot \omega \cdot \beta_{O_2/DN} \cdot C_{NO_3} \quad (7)$$

where K_{BOD} is biochemical degradation rate (day^{-1}), $\beta_{O_2/DN}$ is the yield factor describing the amount of oxygen used for denitrification (gO_2/gN).
   7) Dissolved oxygen (gO_2/m^3):

$$\frac{d(\omega \cdot C_{O_2})}{dt} + \frac{d(Q \cdot C_{O_2})}{dx} = \omega \cdot (K_{BOD} \cdot C_{org} - K_{12} \cdot \beta_{O_2/NT} \cdot C_{NH_4} - K_{RO} \cdot C_{O_2}) \quad (8)$$

where $\beta_{O_2/NT}$ is the yield factor describing the amount of oxygen used for nitrification (gO_2/gN), K_{RO} is reaeration rate (day^{-1}).
   The system of differential equations are approximated by numerical equations with time-space grid $(t_n; x_i) : t_{n+1} = t_n + \tau (n = \overline{0, N})$, $x_{i+1} = x_i + \Delta (i = \overline{1, L})$, where $\tau = const$ is time step, $\Delta = const$ is space step. The upwind approximation scheme is used to solve these equations. It is explicit scheme based on three-point grid [11] [14].
   To remain a characteristic line the model must be calculated using the assumption, where space step must be greater than time step and equation (2) (8) fully describes the river quality behaviour.

4   Parameter Estimation

Almost the entire algal population is composed of N-limited species. Its interaction with dissolved inorganic nitrogen is indirectly described by the nitrification and denitrification coefficients.
   The relatively simple model structure is partially offset by structuring some parameter as a function of the varying river morphology. The model can produce reliable results together with the step-wise parameter variations, only if it is supported by a robust estimation procedure.
   Coefficient values can be obtained in four ways: direct measurement, estimation from field data, literature values, model calibration. Model calibration is usually required regardless of the approach selected. Various predictive equations are used to estimate some coefficients, for example reaeration rate.

The most of popular theoretical formulas show the degree of uncertainty exceeding the degree of forecasted water quality [15].

Indeed, deviations occurred in the BOD and oxygen values which can be attributed mainly to the higher complexity of phenomena involved for such variables. Indeed, ammonia and oxygen concentration values are the results of several chemical, physical and biological processes (i.e., nitrification, denitrification, photosynthesis, atmospheric reaeration, etc.). A slight miscalculation of these processes may contribute to high disagreement between measured and simulated values for BOD and oxygen concentrations. Direct field measurement is the preferred approach for obtaining model input data [10] [16].

In this work the reaeration and biochemical degradation rates are calculated with using annual variation of BOD and oxygen concentrations. The rates were calculated with some assumption. We suppose that organic nutrients reduction is equivalent oxidation reaction. It leads to decreasing of dissolved oxygen concentration. The rate K_{BOD} is defined from the equation (7) without concentration of nitrogen ammonium.

The reaeration rate depends on hydrological conditions of a river. It depends also on oxygen concentration defined by concentration of organic compounds. The reaeration rate K_{RO} is defined from the equation (8). It includes decreasing of dissolved oxygen concentration, biochemical degradation, reaeration, but it doesn't take into account the concentration of nitrate nitrogen.

The equations (7) - (8) are simplified with help of above described assumptions. These equations were approximated by numerical equations with upwind scheme. The biochemical degradation and reaeration rates are defined as in the following:

$$K_{BOD} = \frac{1}{C_{(org)i}^{n+1}} \cdot \left(-\frac{C_{(org)i}^{n+1} - C_{(org)i}^n}{\tau} - \frac{1}{\omega} \cdot \frac{Q_i^{n+1} \cdot C_{(org)i}^{n+1} - Q_{i-1}^{n+1} \cdot C_{(org)i-1}^{n+1}}{\triangle} \right)$$

(9)

$$K_{RO} = \frac{1}{C_{(O_2)i}^{n+1}} \cdot \left(-\frac{C_{(O_2)i}^{n+1} - C_{(O_2)i}^n}{\tau} - \frac{1}{\omega} \cdot \frac{Q_i^{n+1} \cdot C_{(O_2)i}^{n+1} - Q_{i-1}^{n+1} \cdot C_{(O_2)i-1}^{n+1}}{\triangle} + K_{BOD} \cdot C_{(org)i}^n \right)$$

(10)

Space and time grid is used the same as for basic system equations (2) - (8). We used literature values for the transformation rates of nitrogen and phosphorus in this stage of our research [6]. These rates include also temperature influence on eutrophication processes.

5   Results and Discussion

The reaeration and biochemical degradation rates were calculated by means of (9) - (10) equations with using water quality data storage from state monitoring stations. These data were used for annual dynamics estimation of model rates (Fig.1). To specify diurnal dynamics of these rates we used concentrations sampled with short time step. These parameters were measured three times in week during spring-summer period (2013-2015) and every hour during five days in 2014, July.

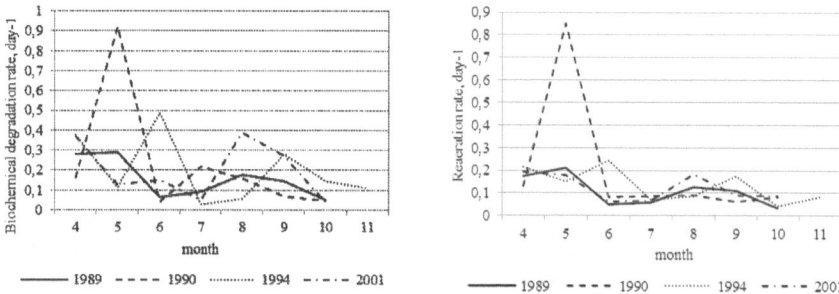

Fig. 1. Annual dynamics of model rates

The mean of reaeration rate varies in the range of 0,01 - 0,3, the mean of biochemical degradation rate varies in the range: 0,1 - 0,5. These rates depend on water pollution levels, varying of river flow rate, seasonal variability of temperature. The river is characterised by a very variable concentration of pollutants. In spring, the photosynthetic activity is considerable, given the high nutrient content of the incoming water, and self-purification is very active. All this factors define annual dynamics of both rates.

The reaeration rate differs in different years, because climate parameters vary also in those years. Seasonally high flow normally occurs during the spring and early period of summer from snowmelt and rains, while seasonally low flow typically occurs during the warmer summer and early fall drought periods. Summer is typically the critical periods for evaluating the worstcase impact of pollutant loads on water quality caused by these seasonal hydrologic and climatological patterns of low flow, minimum dilution, and high temperature. Self-purification intensity is minimal in this period.

Figure 2 shows the comparison between measured and simulated values. The model generally shows a satisfactory capability in reproducing the measured values of nitrogen and phosphorus concentrations. But the calculated BOD values are not so appropriate to measured concentrations. Probably, it can be explained that some parameters of our model are constant, for example, nitrification, denitrification and mineralization rates. We are going to change it in further research.

Eutrophication is difficult processes to exactly mathematical describe because it is quite sensitive to natural environmental conditions. These conditions include physical characteristics such as stream flow, velocity, time of travel, and temperature and chemical/biological characteristics such as in-place sediment oxygen

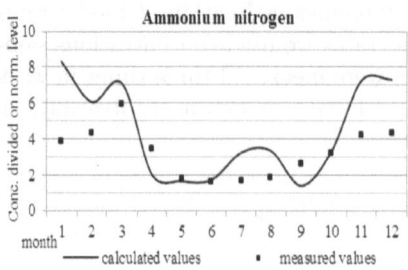
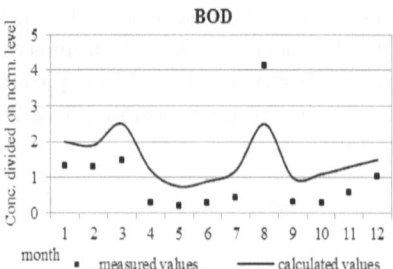

Fig. 2. Calculated and measured values of some model parameters for Kacha river

demand, algal photosynthesis and respiration, and nitrification. The determination of the rates at which various water quality reactions take place in the receiving waterbody introduces additional complications in establishing cause-and-effect relationships and projecting water quality impacts.

Limitations in this model affect the ability to close mass balances, to represent separate biological processes, and to achieve robust model calibration for some parameters. Mass balance problems arise from failure to account for mass in the sediment due to the fundamental imprecision of BOD as a state variable. Further, we suppose to estimate a contribution for all of modeling processes in calculated values. This model turns out to be easy to use and presents interesting perspectives of combination with a simplified hydraulic model to obtain a practical tool.

This work was supported by RFBR grant 15-07-06982

References

1. Bahadur, R., Amstutz, D.E., Samuels, W.B.: Water Contamination Modeling–A Review of the State of the Science. J. of Water Resource and Protection **5**, 142–155 (2013)
2. Li, T., Li, P., Chen, B., Hu, M., Zhang, X.: A Simulation-based Inexact Two-stage Chance Constraint Quadratic Programming for Sustainable Water Quality Management under Dual Uncertainties. J. of Water Resources Planning and Management (2012). doi:10.1061/(ASCE)WR.1943-5452.0000328
3. Shanahan, P., Henze, M., Koncsos, L., Rauch, W., Reichert, P., Somlydy, L., Vanrolleghem, P.: River water quality modelling: II. Problems of the art. In: IAWQ Biennial International Conference, Vancouver, British Columbia, Canada (1998)
4. Toloknova, A.N.: Razrabotka kontseptsii opredeleniya samoochischey sposonbosty vodnyh ecosistem i ee apparaturnaya realizatsiya (2012). http://zhurnal.ape.relarn.ru/articles/2006/107.pdf
5. Vladimirov, A.M., Yu, I.L., Matveev, L.T., Orlov, V.G.: Ohrana okrugaiyuschey sredy. Hydrometeoizdat, Leningrad (1991)
6. Mannina, G., Viviani, G.: River water quality assessment: a hydrodynamic water quality model for propagation of pollutants. J. Water Sci. Technol. **62**(2), 288–299 (2010)

7. Rauch, W., Henze, M., Koncsos, L., Reichert, P., Shanahan, P., Somlydy, L., Vanrolleghem, P.: River water quality modelling: I. State of the art. J. Wat. Sci. Tech. **38**(11), 237–244 (2002)
8. Benedini, M.: Water quality models for rivers and streams. State of the art and future perspectives. J. European Water **34**, 27–40 (2011)
9. Bashenhaeva, N.V.: O samoochischaiuschey sposonbosti reki Selenga. In: Materialy 3 Vserossiyskogo simposiuma s megdunarodnym uchastiem "Fundamentalniye problemy vody i vodnyh resursov", pp. 14–18. ART, Barnaul (2010)
10. Karaushev, A.V.: Metodicheskiye osnovy otsenki antropogennogo vliyaniya nf kachestvo poverhnosbyh vod. Hydrometeoizdat, Leningrad (1981)
11. Belolipetskiy, V.M., Tugovikov, V.B., Tchay, A.A.: Chislennoe modelirovanie protsessov eftrofirovaniya v nignem niefe protsessa ohladiteliya. J. Computational Technology **2**(2), 5–19 (1997)
12. Yu, T.S.: Trehmernaya matematicheskaya model kachestva vod Dneprovsko-bugskogo priustievogo rayona severo-zapadnoy chasti Chernogo moriya. In: Ekol. bez-peka pribreg. ta shelfovoi zon ta kompleks. vikorist resursiv shelfu, vol. 12, pp. 374–391 (2005)
13. Marsili-Libelli, S., Giusti, E.: Water quality modelling for small river basins. J. Environmental Modelling and Software **23**, 451–463 (2008)
14. Kalitkin N.N.: Chislenniye metody. M.: Nauka (1978)
15. Technical Guidance Manual for Performing Wasteload Allocations, Book II: Streams and Rivers Part 1: Biochemical Oxygen Demand/Dissolved Oxygen and Nutri-ents/Eutrophication. EPA document number: EPA-823-B-97-002 (1997)
16. Goncharov, V., Zaslavskaya, B., Isaev, V., Lobchenko, Nichiporova, I.P., et al.: kislorodniy regim rek kak pokazatel produktsionno-destuktsionnyh protsessov v rekah. In: Materialy 5 Vserossiyskogo simposiuma s megdunarodnym uchastiem Organicheskoe veschestvo i niogenniye elementy vo vnutrennih vodoemah i morskyh vodah, pp. 216–218. Kareliskiy nauchniy tsentr RAN, Petrozavodsk (2012) (in Russian)

Modeling of Three-Phase Non-isothermal Flow in Porous Media Using the Approach of Reduced Pressure

N.M. Temirbekov and D.R. Baigereyev$^{(\boxtimes)}$

D. Serikbayev East Kazakhstan State Technical University,
Ust-Kamenogorsk, Kazakhstan
temirbekov@rambler.ru, dbaigereyev@gmail.com

Abstract. This paper focuses on modeling of three-phase non-isothermal compressible flow in porous media taking into account capillary effects. A new formulation of the three-phase non-isothermal flow problem using the concept of reduced pressure is proposed. The purpose of the work is to eliminate the gradients of capillary pressure functions, leading to the unbounded growth of the solution, from the equations for temperature and pressure. An algorithm for the numerical implementation of the model based on the finite-difference method is suggested. A study of the developed difference scheme using the method of a priori estimates is conducted. The simulation results on the example of a one-dimensional model problem are presented.

Keywords: Three-phase non-isothermal flow · Reduced pressure · A priori estimates · Numerical results

1 Introduction

The most common approach to the numerical solution of the three-phase non-isothermal flow problems is based on the selection of pressure of one of the phases, temperature and saturations as the unknowns [1–4]. However, the choice of a phase pressure as the primary variable assumes certain difficulties encountered in the numerical solution of three-phase non-isothermal flow problems, which takes into account capillary effects. Some of them, in relation to the isothermal case, are described in [5–7]. These difficulties are mainly related to the unbounded increase in the derivative of capillary pressure functions when saturations approach corresponding residual values.

To get rid of some of these shortcomings in the numerical solution of the three-phase isothermal flow problems, so-called *global (reduced) pressure - saturations* formulation is widely used. This approach was first proposed in [8] for modeling of isothermal two-phase flow, and then generalized to the isothermal three-phase case. The idea of the reduced pressure approach is to replace the three-phase flow with the flow of some fluid which motion is described by the Darcy's law.

© Springer International Publishing Switzerland 2015
N. Danaev et al. (Eds.): CITech 2015, CCIS 549, pp. 166–176, 2015.
DOI: 10.1007/978-3-319-25058-8_17

In this paper, the idea of introducing the reduced pressure is generalized for the numerical solution of three-phase non-isothermal flow problems. The purpose of this work is to eliminate the gradients of capillary pressure functions, leading to the unlimited growth of the solution from the equations for temperature and pressure through the introduction of a replacement of variables for the pressure. In this paper, the sought substitution is called *the reduced pressure*. A new formulation of the problem, which consists of a system of four partial differential equations with respect to the reduced pressure, temperature and the two phase saturations is proposed. We propose a computational algorithm for the numerical implementation of the model using the finite difference method. A study of the developed finite difference scheme using the method of a priori estimates is conducted. In conclusion, the results of modeling on the example of a one-dimensional model problem are presented.

2   The Derivation of the Model

Let us describe the mathematical model of a three-phase non-isothermal compressible flow in porous media taking into account capillary effects. It is assumed that the movement of phases obeys the generalized Darcy's law. We assume that the phases are in the local thermal equilibrium, so that in any elementary volume the fluids saturating the porous medium and the rock have the same temperature. Furthermore, oil is assumed to be homogeneous non-evaporable fluid and oil reservoir consists of one type of rock. In this case, three-phase non-isothermal flow in a bounded domain $\mathcal{D} \subset \mathbb{R}^d$ ($d = 1, 2, 3$) taking into account capillary forces and the phase transitions between the phases of water and heat transfer is described by the following system of equations:

$$\phi \frac{\partial}{\partial t} (\rho_\alpha s_\alpha) + \nabla \cdot (\rho_\alpha \boldsymbol{u}_\alpha) + \Im_\alpha = q_\alpha, \quad \alpha = w, o, g, \tag{1}$$

$$\boldsymbol{u}_\alpha = -\frac{k k_\alpha}{\mu_\alpha} \nabla p_\alpha, \quad \alpha = w, o, g, \tag{2}$$

$$\frac{\partial}{\partial t} \left(\phi \sum_\alpha \rho_\alpha s_\alpha i_\alpha + (1 - \phi) \rho_r i_r \right) + \nabla \cdot \sum_\alpha \rho_\alpha \boldsymbol{u}_\alpha i_\alpha - \nabla \cdot (k_T \nabla T) = q_T \tag{3}$$

where the subscripts w, o, g, r denote the phases of water, oil, heat transfer, and rock, respectively; ϕ and k are the porosity and permeability of the medium; p_α, s_α, ρ_α, k_α, μ_α, i_α are the pressure, saturation, density, relative permeability, viscosity, and enthalpy of the phase α, respectively; k_T is the coefficient of thermal conductivity; q_α and q_T are source/sink terms and heat flow rate; $\boldsymbol{u}_\alpha$ is the velocity of the phase α, and $\Im_\alpha$ is the rate of phase transitions. Time t is changed in the segment $[0, t_1]$. Additionally, we have the following algebraic constraints:

$$s_w + s_o + s_g = 1, \tag{4}$$

$$p_{ow} = p_o - p_w, \quad p_{go} = p_g - p_o \tag{5}$$

where the capillary pressure functions p_{ow} and p_{go} depend on saturations and temperature and they are assumed to be known. In this work, following [9–12], we neglect the effect of temperature on the capillary pressures.

Finally, the system of equations (1)-(5) is complemented by the initial conditions

$$T\left(x,0\right) = T_0, \quad p_w\left(x,0\right) = p_0, \quad s_\alpha\left(x,0\right) = s_{\alpha 0}, \quad \alpha = w, o,$$

$$s_g\left(x,0\right) = 0, \quad x \in \mathcal{D} \tag{6}$$

and appropriate boundary conditions depending on the particular flow problem.

To derive the model, we first introduce the vector

$$\boldsymbol{u} = \rho_w c_w \boldsymbol{u}_w + \rho_o c_o \boldsymbol{u}_o + \rho_g c_g \boldsymbol{u}_g \tag{7}$$

where c_α is the specific heat of the phase α. Using (2) and (5), one can easily show that

$$\boldsymbol{u} = -k\lambda\left(\nabla p_o - \theta_w \nabla p_{ow} + \theta_g \nabla p_{go}\right) \tag{8}$$

where

$$\theta_\alpha\left(s_w, s_g, p_o, T\right) = \lambda_\alpha \lambda^{-1}, \quad \lambda_\alpha\left(s_w, s_g, p_o, T\right) = \rho_\alpha c_\alpha k_\alpha \mu_\alpha^{-1},$$

$$\lambda\left(s_w, s_g, p_o, T\right) = \lambda_w + \lambda_o + \lambda_g. \tag{9}$$

The main idea of the proposed method is to find a function $p = p\left(s_w, s_g, p_o, T\right)$ which is determined from the differential equation

$$\nabla p = \nabla p_o - \theta_w \nabla p_{ow} + \theta_g \nabla p_{go}. \tag{10}$$

In order to eliminate the gradients of capillary pressure functions ∇p_{ow} and ∇p_{go}, we will look for a function $p_c = p_c\left(s_w, s_g, p, T\right)$ such that

$$\nabla p_c = -\theta_w \nabla p_{ow} + \theta_g \nabla p_{go} + \frac{\partial p_c}{\partial p} \nabla p + \frac{\partial p_c}{\partial T} \nabla T. \tag{11}$$

This holds if and only if the following conditions are met:

$$\frac{\partial p_c}{\partial s_w} = -\theta_w \frac{\partial p_{ow}}{\partial s_w} + \theta_g \frac{\partial p_{go}}{\partial s_w}, \quad \frac{\partial p_c}{\partial s_g} = -\theta_w \frac{\partial p_{ow}}{\partial s_g} + \theta_g \frac{\partial p_{go}}{\partial s_g}. \tag{12}$$

A necessary and sufficient condition for the existence of a function p_c satisfying the conditions (12) is the equality of the mixed derivatives which leads to the condition

$$-\frac{\partial \theta_w}{\partial s_g} \frac{\partial p_{ow}}{\partial s_w} + \frac{\partial \theta_g}{\partial s_g} \frac{\partial p_{go}}{\partial s_w} = -\frac{\partial \theta_w}{\partial s_w} \frac{\partial p_{ow}}{\partial s_g} + \frac{\partial \theta_g}{\partial s_w} \frac{\partial p_{go}}{\partial s_g}. \tag{13}$$

Obviously, the condition (13) limits the choice of functions p_{ow}, p_{go}, k_α, ρ_α and μ_α. When the condition (13) is satisfied, the function p_c is defined as [8]

$$p_c\left(s_w, s_g, p, T\right) = \int_1^{s_w}\left[-\theta_w\left(\eta, 0, p, T\right)\frac{\partial p_{ow}}{\partial s_w}\left(\eta, 0\right) + \theta_g\left(\eta, 0, p, T\right)\frac{\partial p_{go}}{\partial s_w}\left(\eta, 0\right)\right]d\eta +$$

$$+ \int_0^{s_g}\left[-\theta_w\left(s_w, \eta, p, T\right)\frac{\partial p_{ow}}{\partial s_g}\left(s_1, \eta\right) + \theta_g\left(s_w, \eta, p, T\right)\frac{\partial p_{go}}{\partial s_g}\left(s_w, \eta\right)\right]d\eta \quad (14)$$

where p and T are considered as parameters. Now, we can define the sought reduced pressure p in the form

$$p = p_o + p_c. \quad (15)$$

Using (8), (10), (2) and (5), one can easily show that

$$\boldsymbol{u} = -k\lambda\left(\gamma\nabla p - \xi\nabla T\right), \quad (16)$$

$$\boldsymbol{u}_w = \frac{\theta_w}{\rho_w\gamma}\boldsymbol{u} - \frac{k\lambda_w}{\rho_w\gamma}\left(\xi\nabla T - \gamma\left(\nabla p_c + \nabla p_{ow}\right)\right), \quad (17)$$

$$\boldsymbol{u}_g = \frac{\theta_g}{\rho_g\gamma}\boldsymbol{u} - \frac{k\lambda_g}{\rho_g\gamma}\left(\xi\nabla T - \gamma\left(\nabla p_c - \nabla p_{go}\right)\right) \quad (18)$$

where $\xi = \frac{\partial p_c}{\partial T}$, $\gamma = 1 - \frac{\partial p_c}{\partial p}$.

Using the relations (15), (16)-(18), the equations (1)-(5) reduce to the following system of equations for the reduced pressure p, saturations s_w, s_g, and temperature T:

$$a_1\frac{\partial p}{\partial t} + b_1\frac{\partial T}{\partial t} - \nabla\cdot\left(a_2\nabla p\right) + \nabla\cdot\left(b_2\nabla T\right) = Q, \quad (19)$$

$$\phi\frac{\partial s_\alpha}{\partial t} + \phi s_\alpha\left(\beta_{p_\alpha}\frac{\partial p}{\partial t} + \beta_{T_\alpha}\frac{\partial T}{\partial t}\right) + \frac{1}{\rho_\alpha}\nabla\cdot\left(\rho_\alpha\boldsymbol{u}_\alpha\right) + \frac{\mathfrak{I}_\alpha}{\rho_\alpha} = \frac{q_\alpha}{\rho_\alpha}, \quad \alpha = w, g, \quad (20)$$

$$a_3\frac{\partial T}{\partial t} + \boldsymbol{u}\cdot\nabla T - \nabla\cdot\left(k_T\nabla T\right) = Q_T \quad (21)$$

where

$$a_1\left(s_w, s_g, p, T\right) = \sum_\alpha\phi c_\alpha s_\alpha\rho_\alpha\beta_{p_\alpha}, \quad b_1\left(s_w, s_g, p, T\right) = \sum_\alpha\phi c_\alpha s_\alpha\rho_\alpha\beta_{T_\alpha},$$

$$a_2\left(s_w, s_g, p, T\right) = k\lambda\gamma, \quad b_2\left(s_w, s_g, p, T\right) = k\lambda\xi, \quad (22)$$

$$a_3\left(s_w, s_g, p, T\right) = \phi\sum_\alpha\rho_\alpha s_\alpha c_\alpha + (1 - \phi)\rho_r c_r,$$

$$Q = \sum_\alpha c_\alpha\left(q_\alpha \quad \mathfrak{I}_\alpha\right), \quad Q_T = q_T - \sum_\alpha q_\alpha\imath_\alpha,$$

$$\beta_{p_\alpha} = \frac{1}{\rho_\alpha}\frac{\partial\rho_\alpha}{\partial p}, \quad \beta_{T_\alpha} = \frac{1}{\rho_\alpha}\frac{\partial\rho_\alpha}{\partial T}, \quad s_o = 1 - s_w - s_g.$$

To determine the initial and boundary values for the reduced pressure, (15) is used. The initial and boundary conditions for the unknowns s_w, s_g and T remain unchanged.

3 Numerical Implementation of the Model

Now, to test the adequacy of the model, we consider a one-dimensional problem of the displacement of oil by steam on the segment $\mathcal{D} = [0, 1]$, at the ends of which injection and production wells are placed. For the numerical integration of the equations (19)-(21), we use the finite difference method. Let us define a uniform difference mesh $\{x_i = i\Delta x,\ i = 0, 1, ..., N_x\}$ in the domain $\mathcal{D}$, and $t^n = n\Delta t,\ n = 0, 1, ..., M,\ t^0 = 0,\ t^M = t_1$ for the time segment $[0, t_1]$. We associate the following difference problem with the boundary value problem (19)-(21): it is required to find the grid functions $p,\ T,\ s_w,\ s_g$ satisfying the system of difference equations

$$a_1^n p_{\bar{t}}^{n+1} + b_1^n T_{\bar{t}}^{n+1} - \left(a_2^n p_{\bar{x}}^{n+1}\right)_x + \left(b_2 T_{\bar{x}}^{n+1}\right)_x = Q, \tag{23}$$

$$\phi \rho_\alpha^{n+1} s_{\alpha,\bar{t}}^{n+1} + \phi \rho_\alpha^{n+1} s_\alpha^n \left(\beta_{p_\alpha} p_{\bar{t}}^{n+1} + \beta_{T_\alpha} T_{\bar{t}}^{n+1}\right) + \left(\rho_\alpha^{n+1} u_{\alpha x}^{n+1}\right)_{\bar{x}} = q_\alpha^{n+1}, \quad \alpha = w, g, \tag{24}$$

$$a_3^n T_{\bar{t}}^{n+1} - k\lambda \left(\gamma p_x^n - \frac{\partial p_c}{\partial T} T_x^n\right) T_x^{n+1} - \left(k_T T_{\bar{x}}^{n+1}\right)_x = Q_T \tag{25}$$

and the initial and boundary conditions of the form

$$p_i^0 = p_0,\ T_i^0 = T_0,\ s_{w,i}^0 = s_{w0},\quad i = 0, 1, ..., N_x, \tag{26}$$

$$p_0^n = p_1,\ p_{N_x}^n = p_0,\ T_0^n = T_1,\ T_{\bar{x},N_x}^n = 0,$$

$$s_{w,0}^n = s_{w1},\ s_{g,0}^n = s_{g1},\quad n = 1, 2, ..., M. \tag{27}$$

Theorem. Suppose that the coefficients of the system (19)-(21) satisfy the following assumptions:

(A1) There are two positive bounded constants ϕ_m and ϕ_M such that $\phi_m \leq \phi(x) \leq \phi_M$;

(A2) $\lambda_\alpha \in C^0([0, 1]),\ \lambda_\alpha(s_\alpha = 0) = 0$. Furthermore, there are the constants $\lambda_M \geq \lambda_m > 0$ such that

$$0 < \lambda_m \leq \sum_\alpha \lambda_\alpha(s_\alpha) \leq \lambda_M \quad \forall s_\alpha \in [0, 1];$$

(A3) The source/sink terms and heat flow rate q_α, q_T satisfy the conditions $|q_\alpha|_{-1} < \infty,\ |q_T|_{-1} < \infty$;

(A4) The density of the phases are assumed to be $\rho_\alpha \in C^1(\mathbb{R})$ and there are the constants $\beta_m > 0,\ \beta_M > 0$ such that

$$\beta_m < \beta_{p_\alpha} < \beta_M,\quad \beta_m < \beta_{T_\alpha} < \beta_M;$$

(A5) $p_c \in C^1([0, 1])$, furthermore, there are the constants $m_p,\ m_p',$ $m_T,\ m_s > 0$ such that $m_p' < \left|1 - \frac{\partial p_c}{\partial p}\right| \leq m_p,\ \left|\frac{\partial p_c}{\partial T}\right| \leq m_T,\ \left|\frac{\partial p_c}{\partial s_\alpha}\right| \leq m_s$;

(A6) There are the constants $c_M,\ k_m,\ k_M\ k_{T,0}$ such that $k_m \leq k \leq k_M, |c_{p_\alpha}| \leq c_M,\ |c_{V_\alpha}| \leq c_M$ and $|k_T| > k_{T,0}$.

Then, for any initial condition $v^0 = \{p^0, T^0, s_w^0, s_g^0\}$ there exist the numbers M_0, τ_0 depending only on $|Q|_{-1}, |Q_T|_{-1}$ and v^0 such that for any $\tau \leq \tau_0$ and $n \geq 0$, the inequality $E^n \leq M_0$ holds, where

$$E^n = \|p^n\|^2 + \|T^n\|^2 + \|s_w^n\|^2 + \|s_g^n\|^2.$$

Proof. Note that using the assumptions **(A1)-(A6)**, it can be shown that the coefficients of the equations satisfy the inequalities

$$0 < a_{1,m} \leq a_1 \leq a_{1,M}, \quad 0 < b_{1,m} \leq b_1 \leq b_{1,M},$$

$$0 < a_{2,m} \leq a_2 \leq a_{2,M}, \quad 0 < b_{2,m} \leq b_2 \leq b_{2,M}, \quad 0 < a_{3,m} \leq a_3 \leq a_{3,M}$$

for some $a_{k,m}, a_{k,M}, b_{k,m}, b_{k,M} > 0$.

Multiplying the equation (23) by $2\tau p^{n+1}$ and using the assumptions **(A1)-(A6)**, conditions (26), (27) and the formula for summation by parts, we obtain the inequality

$$\left\|p^{n+1}\right\|^2 - \left\|p^n\right\|^2 + \tau^2 \left\|p_{\bar{t}}^{n+1}\right\|^2 + \nu_1 \left\|p_{\bar{x}}^{n+1}]\right\|^2 \leq \frac{2\tau^2 b_{1,M}\varepsilon_1}{a_{1,m}} \left\|T_{\bar{t}}^{n+1}\right\|^2 +$$

$$+ \frac{2\tau^2 b_{2,M}\varepsilon_2}{a_{1,m}} \left\|T_{\bar{x}}^{n+1}]\right\|^2 + \frac{2\tau\varepsilon_3}{a_{1,m}} \|Q\|^2 \tag{28}$$

where positive numbers $\varepsilon_1, \varepsilon_2, \varepsilon_3$ are chosen so that the following inequalities hold:

$$\nu_1 \equiv \frac{1}{a_{1,m}} \left(2a_{2,M}\tau - \frac{b_{1,M}}{16\varepsilon_1} - \frac{b_{2,M}}{2\varepsilon_2} - \frac{\tau}{16\varepsilon_3} \right) > 0, \quad 1 - \frac{2b_{1,M}\varepsilon_1}{a_{1,m}} \geq 0.$$

Similarly, multiplying the equation (25) by $2\tau T^{n+1}$, under the same assumptions, we obtain the inequality

$$\left\|T^{n+1}\right\|^2 - \left\|T^n\right\|^2 + \tau^2 \left\|T_{\bar{t}}^{n+1}\right\|^2 + \frac{2\tau k_T}{a_{3,m}^n} \left\|T_{\bar{x}}^{n+1}]\right\|^2 \leq$$

$$\leq \frac{2\tau^2 \varepsilon_4 k_M \lambda_M m_p}{a_{3,m}^n} \left\|p_{\bar{x}}^n\right\|^2 \left\|T_{\bar{x}}^{n+1}\right\|^2 + \frac{k_M \lambda_M}{2a_{3,m}^n} \left(\frac{m_p}{\varepsilon_4} + \frac{m_T}{\varepsilon_5} \right) \left\|T_{\bar{x}}^{n+1}]\right\|^2 +$$

$$+ \frac{k_M \lambda_M m_p T_0^2}{2a_{3,m}^n \varepsilon_4} + \frac{2\tau^2 \varepsilon_5 k_M \lambda_M m_T}{a_{3,m}^n} \left\|T_{\bar{x}}^n\right\|^2 \left\|T_{\bar{x}}^{n+1}\right\|^2 +$$

$$+ \frac{2\tau\varepsilon_6}{a_{3,m}^n} \|Q_T\|^2 + \frac{\tau}{16a_{3,m}^n \varepsilon_6} \left\|T_{\bar{x}}^{n+1}]\right\|^2. \tag{29}$$

Summing the inequalities (28) and (29), and choosing the numbers $\varepsilon_4, \varepsilon_5, \varepsilon_6$ and the time step τ_0 such that for all $\tau \leq \tau_1$

$$\nu_2' \equiv k_T - \tau k_M \lambda_M \left(\varepsilon_4 m_p \left\|p_{\bar{x}}^n\right\|^2 + \varepsilon_5 m_T \left\|T_{\bar{x}}^n\right\|^2 \right) - \frac{\tau b_{2,M}\varepsilon_2}{a_{1,m}} -$$

$$-\frac{1}{32\varepsilon_6} - \frac{k_M \lambda_M}{2a_{3,m}^n}\left(\frac{m_p}{\varepsilon_4} + \frac{m_T}{\varepsilon_5}\right) > 0,$$

we obtain

$$\left\|p^{n+1}\right\|^2 - \left\|p^n\right\|^2 + \tau^2\left\|p_{\bar{t}}^{n+1}\right\|^2 + \nu_1\left\|p_{\bar{x}}^{n+1}]\right\|^2 + \left\|T^{n+1}\right\|^2 - \left\|T^n\right\|^2 +$$

$$+\tau^2\left(1 - \frac{2b_{1,M}\varepsilon_1}{a_{1,m}}\right)\left\|T_{\bar{t}}^{n+1}\right\|^2 + \nu_2\left\|T_{\bar{x}}^{n+1}]\right\|^2 \le$$

$$\le \nu_3 + \frac{2\tau\varepsilon_3}{a_{1,m}}\|Q\|^2 + \frac{2\tau\varepsilon_6}{a_{3,m}^n}\|Q_T\|^2 \tag{30}$$

where $\nu_2 = \frac{2\tau\nu_2'}{a_{3,m}^n}$, $\nu_3 = \max\left\{\frac{k_M\lambda_M m_p}{2a_{3,m}^n\varepsilon_4}T_0^2, \frac{k_M\lambda_M m_T}{2a_{3,m}^n\varepsilon_5}T_0^2\right\}$. Further, discarding the positive terms on the left side of the last inequality, we have

$$\left\|p_{\bar{x}}^{n+1}]\right\|^2 + \left\|T_{\bar{x}}^{n+1}]\right\|^2 \le \frac{1}{\nu_4}\left(\|p^n\|^2 + \|T^n\|^2 + \nu_3 + \tau\nu_5\|Q\|^2 + \tau\nu_5\|Q_T\|^2\right) \tag{31}$$

where $\nu_4 = \min\{\nu_1, \nu_2\}$, $\nu_5 = \max\left\{\frac{2\varepsilon_3}{a_{1,m}}, \frac{2\varepsilon_6}{a_{3,m}^n}\right\}$.

Using the inequality (31), one can easily show that

$$\left\|u_x^{n+1}\right\|^2 \le \frac{k_M\lambda_M m_0}{\nu_4}\left(\|p^n\|^2 + \|T^n\|^2 + \nu_3 + \tau\nu_5\|Q\|^2 + \tau\nu_5\|Q_T\|^2\right) \tag{32}$$

where $m_0 = \max\left\{m_p^2, m_T^2\right\}$. Using the estimate (32), we have for $\alpha = w, g$

$$\left\|u_{\alpha x}^{n+1}\right\|^2 \le \frac{\nu_6^2}{\nu_4^2}\left(\|p^n\|^2 + \|T^n\|^2 + \nu_3 + \tau\nu_5\|Q\|^2 + \tau\nu_5\|Q_T\|^2\right) +$$

$$+ \nu_6^2\left(\left\|T_{\bar{x}}^{n+1}\right\|^2 + (\gamma_x + \gamma_{cx})^2\right) \tag{33}$$

where $\nu_6 = \frac{k_M\lambda_M m_0}{\nu_4}$.

Similarly, multiplying the equation (24) by $2\tau s_\alpha^{n+1}$, evaluating the scalar products using the assumptions above and the inequality (33), we obtain the inequalities for $\alpha = w, g$. Multiplying the inequalities by the numbers η_w and η_g, respectively, and summing them with the inequality (30), we obtain:

$$\left\|p^{n+1}\right\|^2 - \|p^n\|^2 + \tau^2\nu_9\left\|p_{\bar{t}}^{n+1}\right\|^2 + \nu_{10}\left\|p_{\bar{x}}^{n+1}]\right\|^2 +$$

$$+ \left\|T^{n+1}\right\|^2 - \|T^n\|^2 + \tau^2\nu_{11}\left\|T_{\bar{t}}^{n+1}\right\|^2 +$$

$$+ \sum_{\alpha=w,g}\nu_{12}\left\|T_{\bar{x}}^{n+1}]\right\|^2 + \sum_{\alpha=w,g}\eta_\alpha\left(\|s_\alpha^{n+1}\|^2 - \|s_\alpha^n\|^2 + \tau^2\left\|s_{\alpha,\bar{t}}^{n+1}\right\|^2\right) \le$$

$$\le \nu_3 + \sum_{\alpha=w,g}\frac{2\eta_\alpha\tau}{\phi_m\rho_m}\delta_3\left\|q_\alpha^{n+1}\right\|^2 + \tau\nu_5\|Q\|^2 + \tau\nu_5\|Q_T\|^2$$

where the numbers ε_7, ε_8 and τ_2 were chosen such that for all $\tau < \tau_2$

$$1 - \frac{2\phi_M \rho_M \beta_{p,M} \eta_\alpha \varepsilon_7}{\phi_m \rho_m} - \frac{\tau \nu_7}{\phi_m \rho_m} \sum_{\alpha=w,g} \|s_\alpha^n\|^2 > 0, \quad \alpha = w, g,$$

$$1 - \frac{2b_{1,M} \varepsilon_1}{a_{1,m}} - \frac{2\phi_M \rho_M \beta_{T,M} \eta_\alpha \varepsilon_8}{\phi_m \rho_m} - \frac{\tau \nu_7}{\phi_m \rho_m} \sum_{\alpha=w,g} \|s_\alpha^n\|^2 > 0, \quad \alpha = w, g,$$

$$\nu_1 - \frac{2\tau \rho_M \varepsilon_8 \eta_\alpha}{\phi_m \rho_m} \left(\frac{1}{\rho_m m_p'} \frac{k_M \lambda_M m_0}{\nu_3} \right)^2 > 0, \quad \alpha = w, g.$$

Let M be the positive number such that

$$M = \max \left\{ \nu_3, \ \sum_{\alpha=w,g} \frac{2\eta_\alpha \tau}{\phi_m \rho_m} \delta_3 \|q_\alpha^{n+1}\|^2, \ \tau \nu_5 \|Q\|^2, \ \tau \nu_5 \|Q_T\|^2 \right\}.$$

Denoting $\tau_0 = \min \{\tau_1, \tau_2\}$ and discarding the positive terms in the left-hand side, we obtain for all $\tau < \tau_0$:

$$\|p^{n+1}\|^2 + \|T^{n+1}\|^2 + \sum_{\alpha=w,g} \eta_\alpha \|s_\alpha^{n+1}\|^2 \leq$$

$$\leq \|p^n\|^2 + \|T^n\|^2 + \sum_{\alpha=w,g} \eta_\alpha \|s_\alpha^n\|^2 + 4M.$$

Applying this inequality n times, we obtain

$$\|p^{n+1}\|^2 + \|T^{n+1}\|^2 + \sum_{\alpha=w,g} \eta_\alpha \|s_\alpha^{n+1}\|^2 \leq \|p^0\|^2 + \|T^0\|^2 + \sum_{\alpha=w,g} \eta_\alpha \|s_\alpha^0\|^2 + 4nM.$$

Finally, denoting

$$M_0 = \max \left\{ \|p^0\|^2, \ \|T^0\|^2, \ \eta_w \|s_w^0\|^2, \ \eta_g \|s_g^0\|^2, \ 4nM \right\},$$

we arrive at the inequality

$$\|p^n\|^2 + \|T^n\|^2 + \eta_w \|s_w^n\|^2 + \eta_g \|s_g^n\|^2 \leq M_0$$

for all $n > 0$, which ends the proof of the theorem.

4   Numerical Results

Now we present simulation results on an example of a one-dimensional model problem. In this paper, the following values of parameters of the problem are accepted: $p_1 = T_1 = 1$, $p_0 = T_0 = 0$, $s_{w0} = s_{w1} = 0.3$, $s_{g1} = 0.7$, $s_{rw} = 0.29$, $N_x = 10^3$, $\Delta t = 10^{-5}$. Additionally, the following relations for relative permeabilities, capillary pressures are used in the calculations:

$$k_w = s_w, \quad k_o = 1 - s_w - s_g, \quad k_g = s_g, \tag{34}$$

$$p_{ow} = -0.01 \cdot \ln s_w, \quad p_{go} = 0.1 + 0.01 \cdot \ln(0.0004 \cdot s_g). \tag{35}$$

Fig. 1 shows the dynamics of changes in the reduced pressure at regular intervals $t = 0.1, 0.2, ..., 0.8$. Obviously, the greatest intensity of fluids takes place near the injection well, followed by decrease in the flow direction of the heat transfer. In this regard, the zone of vapor and variable temperature zone characterized by the maximum change in the reduced pressure.

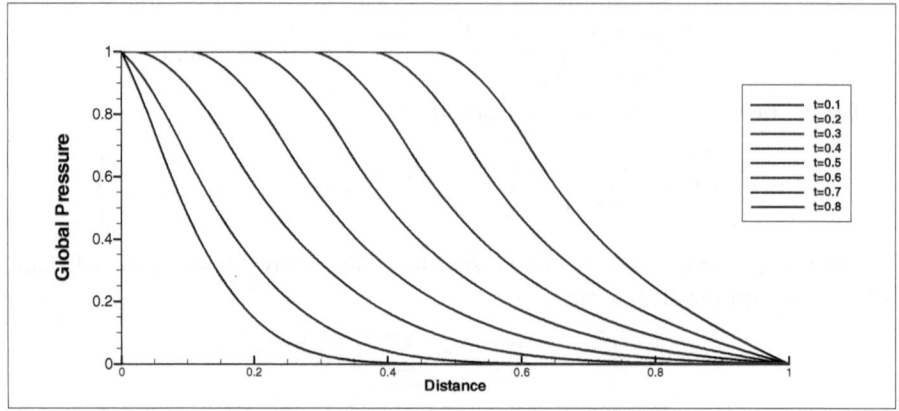

Fig. 1. Distribution of the reduced pressure

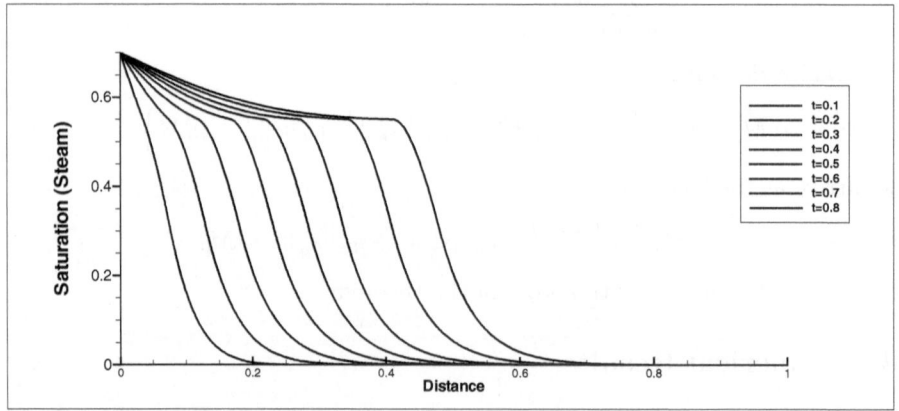

Fig. 2. Distribution of the steam saturation s_g

Figs. 2 and 3 demonstrate the profiles of steam and oil saturations. Near the injection wells, the saturation of steam increases due to evaporation of the water initially filled the reservoir. In the transition zone with variable temperature in the condensation of steam injected, so in the area of water saturation increases dramatically.

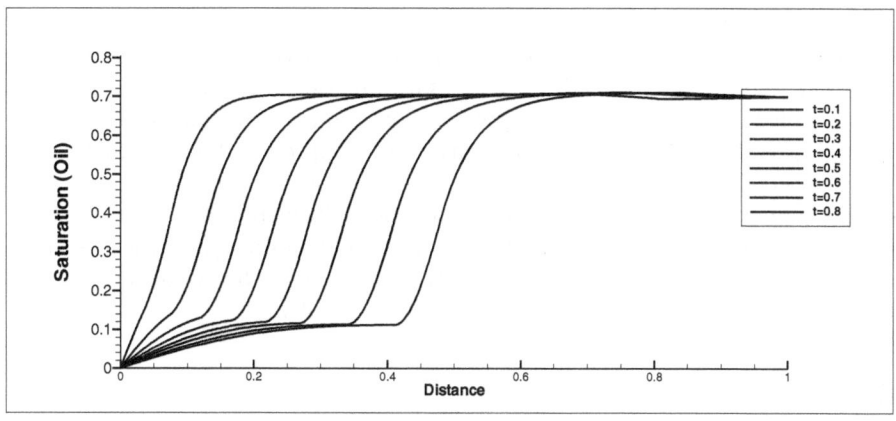

Fig. 3. Distribution of the oil saturation s_o

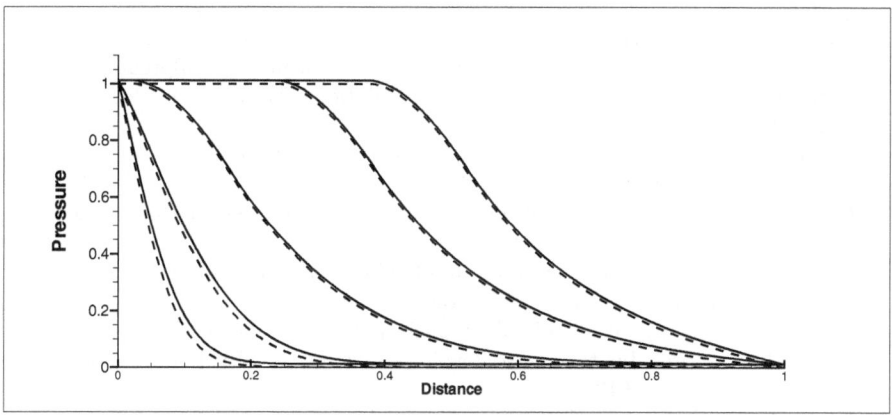

Fig. 4. Distribution of the phase pressure p_o (solid lines) and the reduced pressure p (dashed lines) at $t = 0.03, 0.1, 0.3, 0.55, 0.7$

Figure 4 shows the relation of the reduced pressure p and phase pressure p_o for the times of development $t = 0.03$, $t = 0.1$, $t = 0.3$, $t = 0.55$, $t = 0.7$. The figure shows that the largest deviation occurs in the beginning of the process of oil displacement due to the influence of the capillary pressure gradient ∇p_{ow}. With the passage of the saturation value of water moving away from the values of residual water s_{rw}, and, as a consequence, the deviation decreases.

5   Conclusion

Thus, in this paper, we propose a new formulation of the three-phase non-isothermal flow problem, taking into account the capillary forces and phase transitions between the phases of water and heat transfer. In contrast to the classical formulation of the problem with the use of the phase pressure as the unknown, the gradients of capillary pressure functions are eliminated from the equations for determination of the pressure and temperature which lead to unrestricted growth of the solution when saturations approach their residual values. The conducted study of the difference scheme proves the boundedness of the solution under the condition of boundedness of coefficients in the equations. The results obtained by solving a one-dimensional problem reproduce the characteristics of the process of heat transfer injection that allows the use of the proposed approach to more complex non-isothermal flow problems.

References

1. Chen, Z., Yu, X.: Implementation of mixed methods as finite difference methods and applications to nonisothermal multiphase flow in porous media. Journal of Computational Mathematics **24**(3), 281–294 (2006)
2. Mozzaffari, S., et al.: Numerical modeling of steam injection in heavy oil reservoirs. Fuel **112**, 85–192 (2013)
3. Salimi, H., Wolf, K.H., Bruining, J.: Negative saturation approach for non-isothermal compositional two-phase flow simulations. Transport in Porous Media, 1–22 (2014)
4. Bokserman, A.A., Yakuba, S.I.: Numerical study of the process of steam injection (in Russian). Izvestiya AN SSSR **4**, 78–84 (1987)
5. Chavent, G.: A fully equivalent global pressure formulation for three-phase compressible flow, 19 p. (2009). CoRR, abs/0901.1464
6. Bastian, P.: Numerical Computation of Multiphase Flows in Porous Media. Christian-Albrechts-Universitat Kiel, 236 p. (1999)
7. Saad, B., Saad, M.: Study of full implicit petroleum engineering finite volume scheme for compressible two phase flow in porous media. SIAM Journal of Numerical Analysis, 1–34 (2013)
8. Chavent, G., Jaffre, J.: Mathematical models and finite elements for reservoir simulation, 375 p. Elsevier (1986)
9. Yortsos, Y.C.: Analytical model of oil recovery by steam injection. Ph.D. thesis, California Institute of Technology Pasadena, California, 349 p. (1979)
10. Martinez, M.J., Hopkins, P.L.: LDRD Final Report: Physical Simulation of Non-isothermal Multiphase Multicomponent Flow in Porous Media, Sandia National Laboratories, 65 p. (1997)
11. Gudbjerg, J.: Remediation by steam injection. Ph.D. thesis, Environment & Resources DTU Technical University of Denmark (2003)
12. Adenekan, A.E., Patzek, T.W., Pruess, K.: Modeling of Multiphase Transport of Multicomponent Organic Contaminants and Heat in the Subsurface. Numerical Model Formulation, Water Resources Research **29**(11), 3727–3740 (1993)

Self-Organization Phenomena in Underground Hydrogen Storages

A. Toleukhanov[1]([✉]), M. Panfilov[2], and A. Kaltayev[1]

[1] Al-Farabi Kazakh National University, Almaty, Kazakhstan
amankaznu@gmail.com, aidarkhan.kaltayev@kaznu.kz
[2] University of Lorraine, Nancy, France
michel.panfilov@univ-lorraine.fr

Abstract. The problem of underground hydrogen gas mixture storage is that unlike natural gas, hydrogen gas mixture undergoes chemical changes in underground storage and thus the concentration of hydrogen and carbon dioxide is reduced, and the concentration of methane increases. It has been found that these changes occur because of the activity of methanogenic bacteria populations inhabiting in a reservoir. This chemical activity, which caused by the bacterial activity, as well as gas and water flow in the reservoir causes the phenomenon of self-organization such as the occurrence of autowave spatial structures, the dynamics of which is characterized by a multiplicity of different scenarios, including the occurrence of chaos and the jump from one scenario to another. In this paper we developed a qualitative theory of self-organization scenarios in the underground hydrogen storage depending on the external and internal parameters. Development of the theory and computer models of transport in underground hydrogen storage will be based on the relating of models of multiphase composite flows in porous media with model of dynamics of bacterial populations which will be based on mechanism of chemotaxis (internal chemical mechanism by which bacteria are able to detect the presence of nutrients in the distance and move in that direction).

Keywords: Porous media · Hydrogen · Reactive transport · Bacteria · Methanogenic microorganisms · Population dynamics · Oscillations · Chemotaxis

1 Introduction

Increasing energy demand and anthropogenic greenhouse gas emissions pose serious challenges for national and international energy economies. Low emissions and the increasing efficiency of fuel cells make the case for the use of hydrogen (H_2) as the fuel of the future [1]-[2]. At best, H2 is generated, e.g. through electrolysis, from renewable energy sources. In such a scheme, storing H_2 comes down to storing electricity. However, it may also be produced from fossil fuels, making it easier to contain emissions at the power plants while distributing clean energy in form of H_2, e.g. for transportation.

© Springer International Publishing Switzerland 2015
N. Danaev et al. (Eds.): CITech 2015, CCIS 549, pp. 177–189, 2015.
DOI: 10.1007/978-3-319-25058-8_18

Today underground hydrogen storage (UHS) in aquifers and depleted gas reservoirs is considered as one of the main ways of storing large amounts of energy[3]-[4]. During the last decade it has been found that the behavior of UHS is radically different from the underground storage of natural gas and carbon dioxide, primarily by the fact that in the storage occur chemical changes of hydrogen mixture by present of bacteria in the formations, which absorbs protons of hydrogen, as the energy source. There are several underground hydrogen storages in the UK, USA, Russia, Germany, Czech Republic, Argentine and France. The unusual behavior of hydrogen gas mixture in underground storage has been observed in Lobodice storage of Czech Republic and Baynes of France. By analysis of the gas samples which were taken from the reservoir, it was found that the composition of stored gas has undergone significant changes.

The explanation for these changes lies in the chemical reaction between hydrogen and carbon dioxide, which produce methane and water. In the reservoir conditions it can occur only in the presence of methanogenic bacteria, populations of which have been detected in the derived rock samples[5]-[8]. Thus, the underground storage of hydrogen behaves like a natural chemical reactor, which eventually significantly changes the composition of stored gas. Absorption of the gas components by bacteria leads to intensive growth of the population and increase chemical activity. This chemical activity, which caused by the bacterial activity, as well as gas and water flow in the reservoir causes the phenomenon of self-organization such as the occurrence of autowave spatial structures, the dynamics of which is characterized by a multiplicity of different scenarios, including the occurrence of chaos and bifurcations the jump from one scenario to another. Thus, the new industrial technology - underground storage of electricity in the form of hydrogen - leads to an entirely new scientific issues lying at the intersection of several basic sciences: from hydrodynamics and nonlinear physics to chemistry and microbiology.

The following chemical reaction between injected H_2 and CO_2 occurs in reservoir:

$$CO_2 + 4H_2 = CH_4 + 2H_2O, \ or \ CO + 3H_2 = CH_4 + H_2O \qquad (1)$$

In the present paper we continue to develop the qualitative theory of self-organization in underground hydrogen storage, published first in [9]-[11], for more complicated processes that include two-phase flow and the mechanism of chemotaxis, which is one of the main types of bacterial movement. The analysis is based on the coupled model of two-phase compositional flow and the model of population dynamics.

2   Complete Model of the Process

2.1   Model of Population Dynamics

Let us consider an aquifer which contains an initial population of bacteria, as well as water and gas. Now, mixture of gas in injected where it represents the mixture of H_2 and CO_2 with large domination of hydrogen. Consequently, methanogenic bacteria move to the direction of gas-water contact scince feel nutriments contained in the mixture.

The two-phase system in porous medium represents a fine dispersed alternation of gas bubbles or channels with water channels of droplets. At the macrsocale such a system is considered as two interpenetrating continua coexisting at each space point. The water-gas interfaces which are observed on the pore scale disappear in macroscopic description. At any point two phases are identified by saturation of water S.

Both phases can consist of several chemical components:$(1) = H_2, (2) = CO_2, (3) = H_2O, (4) = CH_4$. The gas phase essentially consists of H_2 and CO_2, while liquid consists mainly of H_2O with low concentration of CO_2, H_2 and CH_4 (the injected gas contains low concentration of CO_2, and hydrogen is low soluble in water). This determines the specific situation when bacteria live in water but the major part of nutriments is concentrated in gas phase.

We consider two kinds of bacteria:

- bacteria present in water: they can be plankton or biofilms attached to pore walls wetted by water;
- the neuston: a biofilm situated just at the interface between water and gas.

Bacteria living in water consume dissolved H_2 and CO_2. Bacteria from neuston consume H_2 and CO_2 directly from the gas phase. On the macroscopic scale (Darcys scale) both phases contain both kinds of bacteria which can be found at any spatial points. Despite the fact that CO_2 in highly soluble in water, it is low present in the injected gas, while hydrogen is very low soluble in water. Therefore, we should consider the concentrations of both these components in water are of the same order.

In gas we have an abundant resource of H_2 and a sufficiently low resource of CO_2. Then the eating rate of bacteria in neuston is controlled only by the concentration of CO_2. Bacterial population can grow due to replication of species and can decay due to natural or forced death. As usually, we assume that the population grow rate is proportional to the eating rate.

Bacteria also can move. We distinguish three types of their motion:

- they can move chaotically similar to brownian motion (bacterial diffusion);
- they can move due to chemotaxis;
- bacteria living in water can be transported by water flow (single-phase bacterial advection);
- bacteria living in neuston can be transported simultaneously with the movement of the water-gas interface (two-phase bacterial advection)

We assume that bacteria in neuston are not transported by chemotaxis but can diffuse. We keep diffusion as it is the mechanism which stabilizes the mathematical properties of the solution, which is considered in the paper [11].

The disappearance of gas-water interfaces in macroscopic equations imposes some difficulties in describing the neuston which represents a pore-scale object. This means that the movement of neuston in macroscopic equations can be obtained by homogenization of its pore-scale motion.

2.2   Balance Equations

Let $n_w(x, t)$ and $n_{ns}(x, t)$ be the number of bacteria per unit volume of porous space in water and in neuston respectively. Taking into account all assumptions formulated above, we can formulate the following equations of population dynamics:

$$\frac{\partial n_{ns}(1-S)}{\partial t} = \eta_{ns}(1-S)\Phi_{ns}(c_g^{(2)}, n_{ns}) - (1-S)\Psi_{ns}(c_g^{(2)}, n_{ns})$$
$$-\langle U_{ns}grad n_{ns}\rangle + div(D_b(1-S)grad n_{ns}) + q_{wn}; \tag{2}$$

$$\frac{\partial n_w S}{\partial t} = \eta_w S\Phi_w(c_w^{(1)}, c_w^{(2)}, n_w) - S\Psi_w(c_w^{(1)}, c_w^{(2)}, n_w) - U_w grad n_w$$
$$+div(D_b S grad n_w) - div(D_{ch}(C^{(1)})Sn_w grad C^{(1)}) - q_w; \tag{3}$$

where subscripts w and ns refer to water and neuston respectively; S is the water saturation; Φ and Ψ are the rate of eating and death of the overall population, their dimension is $mol/(s \cdot m^3)$; η is the rendering coefficient (the coefficient of proportionality between the eating rate and growth rate), its dimension is $1/mol$; q_{wn} is the rate of bacteria transition from water to neuston; D_b is the coefficient of bacterium diffusion in bulk water; U_{ns} is the velocity of movement of gas-liquid interface; U_w is the water flow velocity; $c_i^{(k)}$ is the mole fraction of chemical component k in phase i; $C^{(k)}$ is the total mole fraction of chemical component k in both phases.

Term $\langle U_{ns}grad n_{ns}\rangle$ represent the advective movement of neuston homogenized over an elementary representative volume of porous medium. As mentioned above, the neuston represents a pore-scale object, so the velocity U_{ns} is a pore-scale variable.

The relation between $c_i^{(k)}$ and $C^{(k)}$ is as follows:

$$C^{(k)} = \frac{\rho_w c_w^{(k)} S + \rho_g c_g^{(k)}(1-S)}{\rho_w S + \rho_g(1-S)} \tag{4}$$

where ρ_i is the molar density of phase i (mol/m^3).

Functions Φ_{ns} and Φ_w have the meaning of the number of moles of nutriments consumed by all bacteria during $1s$ in a volume unite. The ratios Φ/n and Ψ/n are the individual rates of eating and decay per one bacterium. The rendering coefficient η determines at what degree the colony growth rate is different from the eating rate. In particular, if $\eta < 1$ then growing is slower than eating.

In general case the individual eating rate Φ/n depends on the size of the population and on the amount of nutriment. These two effects considered in paper [11] in more detail is provided below:

$$\Phi_{ns} = \frac{1}{t_{e,ns}} \frac{n_{ns} c_g^{(2)}}{(1+a_{ns} c_g^{(2)})}, \quad \Phi_w = \frac{1}{t_{e,w}} \frac{n_w^2}{(1+n_w^2/n_{wn}^2)} \frac{c_w^{(1)} c_w^{(2)}}{(1+a_{w1} c_w^{(1)})(1+a_{w2} c_w^{(2)})},$$

$\Psi_{ns} = \frac{n_{ns}}{t_d}, \Psi_w = \frac{n_w}{t_d}$. where $t_{e,ns}$ and $t_{e,w}$ are characteristic time of eating at vanishing resource; t_d is the time of decay; a_{ns}, a_{w1}, a_{w2} are three additional empirical coefficients.

2.3   Reduced Equation of Population Dynamics

Two equations (2),(3) with respect to n_{ns} and n_w may be reformulated with respect to the total number of bacteria $n = n_{ns}(1 - S) + n_w S$ and the ratio $\theta = n_w S/n$. Respectively $n_{ns}(1 - S)/n = 1 - \theta$. Equation for n has the following form:

$$
\begin{aligned}
\frac{\partial n}{\partial t} &= \frac{\eta_{ns} c_g^{(2)}(1-\theta)n}{t_{e,ns}(1+a_{ns}c_g^{(2)})} + \frac{\eta_w c_w^{(1)} c_w^{(2)} \theta^2 Sn^2}{t_{e,w}(S^2+\frac{\theta^2 n^2}{n_{wn}^2})(1+a_{w1}c_w^{(1)})(1+a_{w2}c_w^{(2)})} - \frac{n}{t_d} \\
&+ \langle U_{ns}grad\frac{(1-\theta)n}{(1-S)}\rangle + div(D_b(1 - S)grad\frac{(1-\theta)n}{(1-S)}) - U_w grad\frac{\theta n}{S} \\
&+ div(D_b Sgrad\frac{\theta n}{S}) - div(D_{ch}(C^{(1)})\theta ngradC^{(1)})
\end{aligned}
\tag{5}
$$

It can be simplified.

First of all, it is possible to neglect the neuston advection in the first approximation. Indeed, the flow of water and gas does not mean that the interface between them moves. A movement of the interface means, on the maroscale, that the local saturation changes. Therefore the term $\langle U_{ns}grad...\rangle$ is proportional to $\frac{\partial S}{\partial t}$. For slow variation of saturation in time, this term can be neglected.

The second approximation takes into account the fact that concentrations of CO_2 and H_2 in water are low. This means that terms $a_{w1}c_w^{(1)}, a_{w2}c_w^{(2)}$ and $a_{ns}c_g^{(2)}$ are low with respect to 0.

The third simplification consists of assuming that the value of n_{wn} which corresponds to the state of satiety is high, then the value $\frac{n}{n_{wm}} \to 0$.

The fourth approximation can consist of assuming that the ratio θ is close to water saturation S. This means that the fraction of the number of bacteria in neuston is of the same order as gas saturation, and the fraction of bacterial number in water is of the same order as water saturation. Then we can use only one equation for n to describe population dynamics. It takes the following form:

$$
\begin{aligned}
\frac{\partial n}{\partial t} &= \eta_{ns}(1 - S)\frac{c_g^{(2)}n}{t_{e,ns}} + \eta_w S\frac{c_w^{(1)}c_w^{(2)}n^2}{t_{e,w}} - \frac{n}{t_d} \\
&- U_w gradn + div(D_b gradn) - div(D_{ch}(C^{(1)})SngradC^{(1)})
\end{aligned}
\tag{6}
$$

where $D_{ch}(C^{(1)}) = D_{ch}^{max}e^{-\lambda_{ch}C^{(1)}}$ is the decreasing function, n is bacteria number.

2.4   Reactive Transport of Chemical Components

For reactive multi-component transport, the main problem is the coupling between components through the reactive term. Indeed the reaction kinetics depends on the concentrations of several components, which makes necessary to consider large system of coupled transport equations. For the case of an irreversible reaction, the situation is simplified because the reaction kinetics depends only on reagents and does not depend on the reaction products. Then it is sufficient to formulate the transport equations only for hydrogen and CO_2.

The reaction rate is totally controlled by bacteria and, thus, is equal to the rate of bacterial eating $\Phi_w S + \Phi_{ns}(1 - S)$. According to the formula of the chemical reaction (1), one mole of consumed nutriment contains 1/5 of CO_2 and

4/5 of H_2. As the result, the model of CO_2 and H_2 transport has the following form: $k = 1, 2$

$$\phi\frac{\partial}{\partial t}(\rho_g c_g^{(k)}(1-S) + \rho_w c_w^{(k)}S) + div(\rho_g c_g^{(k)}V_g^{(k)} + \rho_w c_w^{(k)}V_w^{(k)}) =$$
$$\frac{1}{\Omega}G^{inj}c^{(k),inj} - \frac{\phi\gamma^{(k)}(1-S)c_g^{(2)}n}{t_{e,ns}} - \frac{\phi\gamma^{(k)}c_w^{(1)}c_w^{(2)}Sn^2}{t_{e,w}} \qquad (7)$$
$$+div(\rho_g D_g^{(k)}\phi(1-S)gradc_g^{(k)} + \rho_w D_w^{(k)}\phi S gradc_w^{(k)})$$

For the total fluid:

$$\phi\frac{\partial}{\partial t}(\rho_g(1-S) + \rho_w S) + div(\rho_g V_g + \rho_w V_w) = \frac{1}{\Omega}G^{inj} \qquad (8)$$

where
$$V_g = -\lambda_g(gradP_g - \rho_g^m g), \; V_w = -\lambda_w(gradP_w - \rho_w^m g), \; \lambda_i = \frac{Kk_i(S)}{\mu_i};$$
$$V_i^{(k)} = V_i + V_{iD}^{(k)} = -\frac{\phi D_i^{(k)}S_i}{c_i^{(k)}}, \; i = g, w;$$
$$P_w = P_g - P_c(S);$$
Dissolution:

$$c_g^{(3)} = 1 - c_g^{(1)} - c_g^{(2)}; \qquad (9)$$

$$c_w^{(k)} = H^k(P_w)c_g^{(k)}, k = 1, 2, 3; \qquad (10)$$

where S is the water saturation; P is the pressure; ρ is the molar density; μ is the dynamic viscosity; K is the absolute permeability; ϕ is the porosity; $k_{g,w}(S)$ is the relative permeability; G^{inj} is the molar rate of gas injection (mol/s), Ω is the total volume of the reservoir, $c^{(k),inj}$ is the injection concentration of component k in the injected gas (constant value);$\gamma^{(k)} = \begin{cases} 4/5, k = 1 \\ 1/5, k = 2 \end{cases}$. $C^{(k)}$ is the total mole fraction of chemical component k in both phases.

3   Analytical and Numerical Results

3.1   Asymptotic Model for Low Gas Saturation

Let's consider asymptotic model for low gas saturation. In this case the neuston is very modest and bacteria living in water dominate far from the interface. Consequently, the chemotaxis which determines the neuston formation should be taken into account. Since the reaction kinetics depends on concentrations of both CO_2 and H_2, the model of the process resulting from (6) and (7) consists of three equations in this case:

$$\begin{cases} \frac{\partial c_1}{\partial t} = q_1 - \alpha_1 c_1 c_2 n^2 + D_w^{(1)}\Delta c_1 \\ \frac{\partial c_2}{\partial t} = q_2 - \alpha_2 c_1 c_2 n^2 + D_w^{(2)}\Delta c_2 \\ \frac{\partial n}{\partial t} = -\beta n + \alpha_3 c_1 c_2 n^2 + D_b\Delta n - D_{ch}^{max}\nabla(exp(-\lambda_{ch}c_1)n\nabla c_1) \end{cases} \qquad (11)$$

where

$$c_k = c_g^k, \quad q_k = \frac{G^{inj}c^{(k),inj}}{\Omega\phi\rho_w H^{(k)}S}, \quad \alpha_1 = \frac{4H^{(2)}}{5t_{e,w}\rho_w}, \quad \alpha_2 = \frac{4H^{(1)}}{5t_{e,w}\rho_w}, \quad \alpha_3 = \frac{\eta_w H^{(1)}H^{(2)}S}{t_{e,w}},$$

$$\beta = \frac{1}{t_d}$$

Moreover, when the concentration of one of the components is very low, we obtain the model which may be analyzed without simplifications. Let us assume that water can contain very low concentration of hydrogen, that is $c_1 << c_2$. Then concentration c_2 may be considered as variable with small change. From (11) the following expression is obtained:

$$\begin{cases} \frac{\partial c_1}{\partial t} = q_1 - \alpha_1 c_1 n^2 + D_w^{(1)}\Delta c_1 \\ \frac{\partial n}{\partial t} = -\beta n + \alpha_3 c_1 n^2 + D_b\Delta n - D_{ch}^{max}\nabla(exp(-\lambda_{ch}c_1)n\nabla c_1) \end{cases} \tag{12}$$

which is the Turing model [12], if chemotaxis term is neglected.

3.2   Analytical Study: Limit Stationary Spatial Waves

The resulting model (12) has a limit cycle in time, if diffusion term and chemotaxis are neglected. In the paper [10],[13]-[14] there is a criterion for the existence of a limit cycle for the case $\alpha_1 = \alpha_3 = \beta = 1$:

$$0.90032 < q_1 < 1.0 \tag{13}$$

Stationary solutions of the system (12) represent the second kind of limit behavior at $(t \to \infty)$. In the 1D case the system of equations correspond to the model:

$$\begin{cases} D_w^{(1)}\frac{\partial^2 c_1}{\partial x^2} = \alpha_1 c_1 n^2 - q_1 \\ D_b\frac{\partial^2 n}{\partial x^2} - D_{ch}^{max}\frac{d}{dx}(exp(-\lambda_{ch}c_1)n\frac{d}{dx}c_1) = \beta n - \alpha_3 c_1 n^2 \end{cases} \tag{14}$$

The simplest non-trivial stationary solution corresponds to the limit case: $D_{b=0}, D_{ch}^{max} = 0$

$$\begin{cases} D_w^{(1)}\frac{\partial^2 c_1}{\partial x^2} = \alpha_1 c_1 n^2 - q_1 \\ 0 = \beta n - \alpha_3 c_1 n^2 \end{cases} \tag{15}$$

System (15) requires two boundary conditions:

$$c_1\mid_{x=0} = c^0, \frac{dc_1}{dx}\mid_{x=0} = 0 \tag{16}$$

Then the system (15) may be reduced to one ordinary differential equation of the second order:

$$\frac{d^2 c_1}{dx^2} = f(c_1), f(c_1) = \frac{\alpha_1\beta^2}{\alpha_3^2 D_w^{(1)}c_1} - \frac{q_1}{D_w^{(1)}} \tag{17}$$

Equation (17) may be reduced to a non-linear autonomous dynamic system:

$$\begin{cases} \frac{\partial c_1}{\partial x} = U \\ \frac{\partial U}{\partial x} = f(c_1) \end{cases} \tag{18}$$

We have obtained the non-linear autonomous second-order system which can be analysed using the traditional methods of the theory of non-linear dynamics. The Jacobi matrix of the system is:

$$J = \begin{pmatrix} -1 & 1 \\ -\frac{\alpha_3^2 q_1^2}{\alpha_1 D_w^{(1)} \beta^2} & -1 \end{pmatrix} \tag{19}$$

One stationary point exists: $U_s = 0$, $c_s = \frac{\alpha_1 \beta^2}{\alpha_3^2 q_1}$. Then we calculate eigenvalues $\nu_{1,2}$ of matrix J at the stationary point U_s, c_s:

$$det(J - \nu I) = \begin{vmatrix} -\nu & 1 \\ -\frac{\alpha_3^2 q_1^2}{\alpha_1 D_w^{(1)} \beta^2} & -\nu \end{vmatrix} = 0$$

The eigenvalues are:

$$\nu_{1,2} = \pm \frac{\alpha_3 q_1}{\sqrt{\alpha_1 D_w^{(1)} \beta}} i \tag{20}$$

Thus, point $p(U_s, c_s)$ is the center if the expression $\frac{\alpha_3 q_1}{\sqrt{\alpha_1 D_w^{(1)} \beta}} > 0$ is positive, which is the condition of existence of periodic solutions of system (15). The phase portrait of (15) calculated for $\alpha_1 = \alpha_3 = 1$, $D_w^{(1)} = 1$ and $q_1 = 0.95$ is shown in Fig.1. The corresponding periodic oscillations of H_2 concentration are shown in Fig.2 for the case of the boundary condition:

$$c_1 \mid_{x=0} = 0.1, \frac{dc_1}{dx} \mid_{x=0} = 0 \tag{21}$$

3.3   Numerical Study

Then we analyze the problem (12) of gas injection in two-dimensional case with constant initial conditions and Neumann boundary conditions which correspond to impermeable boundaries:

$$n \mid_{t=0} = 1, c_1 \mid_{t=0} = 1, \frac{dc_1}{d\nu} \mid_{\partial \Omega} = 0, \frac{dn}{d\nu} \mid_{\partial \Omega} = 0 \tag{22}$$

The initial values are located within the zone of attraction of the limit cycle, so that the solution of this problem is space-invariant and oscillating in time. The flow rate q_1 in equation (12) represents the hydrogen injection into the reservoir.

This space-invariant solution was perturbed in the form of an instantaneous non-zero concentration gradient applied to the small vicinity of the origin. The evolution of the perturbation is shown in Fig. 3- Fig. 4. And, Table 1 shows the data used in the calculations.

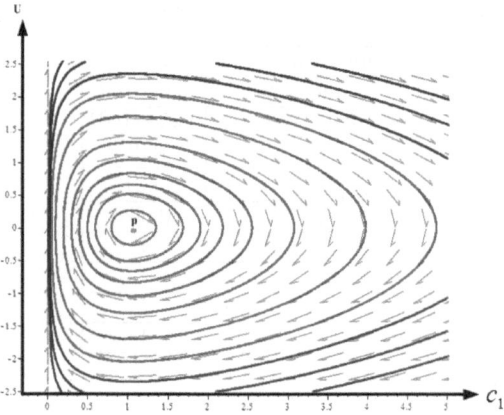

Fig. 1. Phase portrait of stationary system (18)

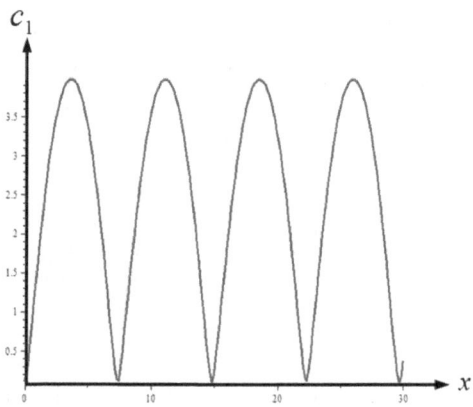

Fig. 2. Stationary periodic behavior of hydrogen concentration

After perturbation, the irregular waves traveling throughout the overall domain were observed. Their evolution was very fast establishing to the structure presented by regular periodic waves invariable in time. The Fig. 3 and 4 represents the results of numerical calculation of the evolution of the hydrogen concentration, changes in the number of bacteria at $t = 40..1000$ with diffusion as well as taking into account chemotaxis which was used the calculated data from Table 1. This means sufficiently regular ring waves are developed with excess and deficiency of hydrogen and bacteria in the space, which alternate with each other. In areas with high bacteria concentrations where the reaction (1) is rapid, alternation with the ring excess and deficiency of bacteria appear, whereby the methanogenic bacteria generates methane.

Table 1. Calculated data.

Computational grid	32×32
Time step	0.006104
q_1 (perturbation)	$0.95 + 0.01$
q_1	0.95
$D_w^{(1)}$	0.01
D_{ch}^{max}	0.001
α_1	1
α_3	1
β	1
λ_{ch}	1

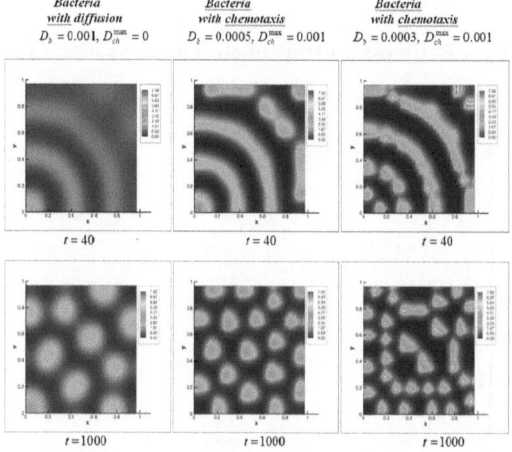

Fig. 3. Evolution of auto-waves of bacterial population with diffusion and chemotaxis at $t = 40..1000$

In case of taking into account the chemotaxis of bacteria, the bacteria forms neuston formation. In this work an attempt has been carried out to qualitatively analyze the impact of methanogenic bacteria on the dynamics of the formation of methane in underground hydrogen storage. Occurrence of undamped oscillations during the time which tends asymptotically to periodic waves, means that the system undergoes self-organization of new structures in the form of methane. It should be noted that, not only in the case of consideration of diffusion but also chemotaxis damping oscillations were observed in space. In the limit of computational time steady-state spatial pattern of frozen waves is observed. Following results in Fig. 2 and Fig. 4 predicts the effect of a natural in situ separation of hydrogen gases, which was observed in practice.

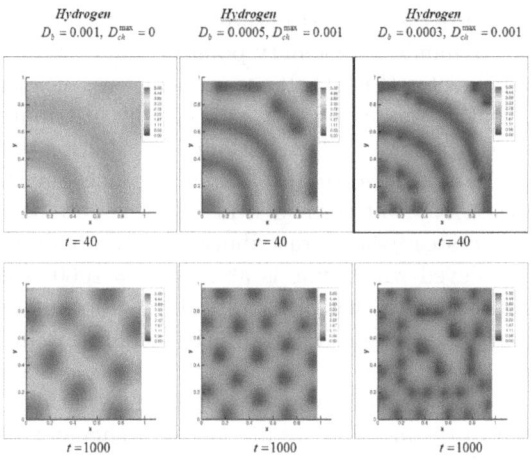

Fig. 4. Hydrogen concentrations at $t = 40..1000$

4 Conclusion

In papers [7] and [8] it was proved that an underground storage of hydrogen can function as a natural chemical reactor producing methane from hydrogen and carbon dioxide. The reaction between H_2 and CO_2 (1) is catalyzed by methanogenic bacteria and happens in the form of the metabolism process.

In paper [10] the first mathematical model of the process was developed. It was based on single-phase flow model coupled with population dynamics equation. The bacterial population was considered in the average and various forms of its existence were reflected in nonlinear kinetics of population growth.

In paper [11] we developed the two-phase flow model coupled with the dynamics of two bacterial populations. One of them represents bacteria living in water, while the second one is the neuston - a thin biofilm situated at the interfaces between water and gas. We also developed the mathematical model of chemotaxis in two-phase fluid, which is the main mechanism of neuston formation: bacteria living in water feel the presence of nutriments concentrated in the injected gas and move to the direction of the interfaces water-gas without crossing them. We have shown that the chemotaxis law should satisfy some specific conditions to ensure the neuston formation. In particular, the bacteria diffusion is shown to be a regularizing mechanism which ensures mathematically the existence of regular solutions.

In this paper, we used above mentioned mathematical model from [11], analystically and numerically studied the equations of mathematical model for the case of low gas saturation, and compared the results.

In case of low gas saturation, asymptotic model (12) related to hydrogen and population of bacteria, taking into account $D_b = 0$ and $D_{ch}^{max} = 0$ when $t \to \infty$, was analytically studied. This model leads to non-linear autonomous dynamic system which depends on space coordinates, that is, (18). When the model in

(18) which is second-order system was studied using traditional methods of the theory of non-linear dynamics, stationary point of that system was found and its type found out to be center (Fig. 1). It means that the considered system has a periodic solution, that is, the hydrogen is periodically distributed over the space which can be noticed from Fig. 2. This periodic distribution phenomenon is also detected, when model (12) is solved by taking into account the bacteria diffusion and chemotaxis.The results provided in Fig. 4 is obtained by perturbation theory. Moreover, the steady-state distribution of the hydrogen concentration over the space is achieved when time is at least $t = 1000$. After $t = 1000$ the periodic distribution is frozen, which means, now, it does not depend on time. The obtained results describe natural gas in situ separation in the underground hydrogen storage.

The results in Fig. 3 compare two cases: a) the model takes into account the diffusion of bacteria b) the model takes into account chemotaxis of bacteria. It is noticed that in case of chemotaxis, amount of bacteria was higher concentrated in the places where hydrogen concentration is high compared to the case when the model uses diffusion of bacteria. This physically means that more the bacteria is concentrated faster the chemical reaction (1). As a result of this phenomena, the methane gas is generated in underground hydrogen storage.

In a word, the results provided above show the natural in situ separation of hydrogen mixture and the generation of methane gas during the underground hydrogen storage.

References

1. Andrews, J., Shabani, B.: Re-envisioning the role of hydrogen in a sustainable energy economy. International Journal of Hydrogen Energy **37**(2), 1184–1203 (2012)
2. Simbeck, D.R.: CO2 capture and storage the essential bridge to the hydrogen economy. Energy **29**, 1633–1641 (2004)
3. Zittel, W., Wurster, R.: Hydrogen in the energy sector. Laudwig-Bolkow Systemtechnik GmbH **7**, August 1996
4. Taylor, J.B., Alderson, J.E.A., Kalyanam, K.M., Lyle, A.B., Phillips, L.A.: Technical and economic assessment of methods for the storage of large quantities of hydrogen. Int. J. Hydrog. Energy **11**(1), 22 (1986)
5. Bulatov G.G., Underground storage of hydrogen. Ph.D. thesis, Moscow Gubkin Oil and Gas University (1979) (in Russian)
6. Carden, P.O., Paterson, L.: Physical, chemical and energy aspects of underground hydrogen storage. International Journal of Hydrogen Energy **4**(6), 559–569 (1979)
7. Buzek, F., Onderka, V., Vancura, P., Wolf, I.: Carbon isotope study of methane production in a town gas storage reservoir. Fuel **73**(5), 747–752 (1994)
8. Smigai, P., Greksak, M., Kozankova, J., Buzek, F., Onderka, V., Wolf, I.: Methanogenic bacteria as a key factor involved in changes of town gas in an underground reservoir. FEMS Microbiol. Ecol. **73**, 221–224 (1990)
9. Panfilov, M., Gravier, G., Fillacier, S.: Underground storage of H2 and H2-CO2-CH4 mixtures. In: Proc. ECMOR-X: 10th European Conference on the Mathematics of Oil Recovery, 4–7 September 2006. Amsterdam, the Netherlands, Ed. EAGE, 2006, paper A003

10. Panfilov, M.: Undeground storage of hydrogen: self-organisation and methane generation. Transport in Porous Media **85**, 841–865 (2010)
11. Toleukhanov, A., Panfilov, M., Panfilova, I., Kaltayev, A.: Bio-reactive two-phase transport and population dynamics in underground storage of hydrogen: natural self-organisation. In: Proc. ECMOR-XIII: 13th European Conference on the Mathematics of Oil Recovery, 10–13 September 2012. Biarritz, France, Ed. EAGE, 2012, paper B09
12. Turing A.M. The chemical basis of morphogenesis. Philos. Trans. R. Soc. London, Ser B **B237**, 37–72 (1952)
13. Merkin, J.H., Needham, D.J., Scott, S.K.: On the creation, growth and extinction of oscillatory solutions for a simple pooled chemical reaction scheme. SIAM J. Appl. Math **47**, 1040–1060 (1987)
14. Schnakenberg, J.: Simple chemical reaction systems with limit cycle behaviour. J. Theor. Biol. **81**(3), 389–400 (1979)

Numerical Modeling of Posteriori Algorithms for the Geophysical Monitoring

Gyulnara Voskoboynikova$^{(\boxtimes)}$ and Marat Khairetdinov$^{(\boxtimes)}$

Institute of Computational Mathematics and Mathematical Geophysics
of Siberian Branch of Russian Academy of Sciences, Lavrentiev Ave, 6,
630090 Novosibirsk, Russia
{gulya,marat}@opg.sscc.ru
http://www.sscc.ru

Abstract. In this paper, some problems of geophysical monitoring of the natural environment are considered. Many of them involve online detection of natural and technogenic events and the preceding geodynamic processes developing in the Earth. Such events include earthquakes, volcano eruptions, lunar and solar tides, landslides, falls of celestial bodies, quarry explosions causing technogenic earthquakes, etc. A new approach to solving the problem of active geophysical monitoring of the natural environment is proposed and investigated. It is based on the detection and separation of waveforms generated in the Earth and surface atmosphere by the above events. The solution is obtained by a unifying process of discrete optimization. The efficiency of this approach is illustrated by some numerical experiments.

Keywords: Geophysical monitoring · Natural and technogenic events · Inverse problems · Posteriori algorithms · Numerical experiments · Seismic location · Borehole source

1 Introduction

At present, the monitoring, prediction, and prevention of natural and technogenic disasters are among priority problems. Many of them are associated with geophysical monitoring of natural and technogenic events and the preceding geodynamic processes. Such events are earthquakes, volcano eruptions, lunar and solar tides, landslides, falls of celestial bodies, quarry explosions causing technogenic earthquakes, etc. Monitoring has several successive stages, including recording of responses to events and measurement of their major parameters, such as travel times of seismic waves or initial waveforms. At the final stage, inverse problems of determining the geographical location and recording time of an event are solved. The problem of determining the geometric parameters of the underground zone of preparation for catastrophic events is even more

G. Voskoboynikova—This work was supported by the Russian Foundation for Basic Research, projects No 14-07-00518, 15-07-10120K.

complicated. A popular method for solving the inverse problems is the least-squares method based on simple calculations. At the same time, this method is sensitive to crude measurement errors (large deviations) in the initial data, which indicates its limited character [1]. Therefore, it is important to increase the accuracy of estimating the wave parameters in noise. In this paper, a new approach is proposed. In comparison to the known methods of statistical data processing, it provides higher accuracy in the measurement of wave arrival times and simultaneous separation of their forms. This approach is based on a posteriori computational algorithms of discrete optimization. The results of numerical experiments on estimating the accuracy and noise immunity of the algorithms are presented.

2   Methods for Solving the Problem of Geophysical Monitoring

2.1   Problem Statement

The problem of estimating unknown parameters of an event is reduced to solving the nonlinear system of equations

$$\hat{\eta} = \eta(\gamma, \theta) + \varepsilon, \tag{1}$$

where $\hat{\eta} = (\hat{n}_1, ..., \hat{n}_N)^T$ is the vector of measured wave's travel times, $\eta(\gamma, \theta) = (n_1, ..., n_N)^T$ is the N-dimensional vector of calculated travel times (theoretical hodograph) or the regression function, $\varepsilon = (\varepsilon_1, ..., \varepsilon_N)^T$ is the residual vector, $\theta = (x, y, z, v, t)^T$ is the m-dimensional vector of estimated parameters, $\gamma = (\gamma_1, ..., \gamma_N)$ is the matrix of the sensors coordinates, and N is the sensors number. The space coordinates of the source x, y, z are the parameters to be estimated, v is the velocity in the medium, and t is the time in the source. The parameters are estimated using information about the distribution of the errors $\varepsilon_i = \hat{\eta}_i(x_i, \theta) - \eta_i(x_i, \theta), i = 1, ,, ., N$. In what follows, it will be assumed that $\varepsilon_i, i = 1, ..., N$ are mutually independent random quantities. At small values of $\varepsilon_i, i = 1, ..., N$ (as in the case of a sufficiently dense observation system) and without "large deviations", many distributions of $\varepsilon_i, i = 1, ..., N$ in the limiting case transform to a normal distribution with zero mean and given variances: $E\varepsilon_i = 0, E\varepsilon_i\varepsilon_j = \sigma_i^2\delta_{ij}, \sigma_i = \sigma(x_i)$, where δ_{ij} is the Kronecker symbol, $i = 1, ..., N$.

The solution to equation (1) is reduced to solving the inverse problem. In this case, the accuracy of the solution is in estimating errors of the time vector $\hat{\eta}$, characterized by the variance σ_η^2, the errors $\varepsilon = (\varepsilon_1, ..., \varepsilon_N)^T$, and choosing sensor arrangement geometry on the Earth's daily surface. In particular, for a triad of seismic stations $(N = 3)$ errors in determining the azimuth to the source and the "source - receiver" distance in the polar system of coordinates are, respectively,

$$\sigma_{AZ}^2 = \sigma_\eta^2 F_1(\bar{\eta}, \bar{\gamma}), \sigma_R^2 = \sigma_\eta^2 F_2(\bar{\eta}, \bar{\gamma}). \tag{2}$$

Here σ_η^2 is the estimation error of the travel times, and $F_1(\bar{\eta}, \bar{\gamma})$, $F_2(\bar{\eta}, \bar{\gamma})$ are the functions depending on geometry of arrangement of gauges, their positions concerning a source and a vector of a transit time of waves. The expressions (2) show that the error in determining the source coordinates directly depends on the errors in measurement of the travel times, which calls for their minimization. The major stages of solving the problem are as follows: 1) detection of waves on the background of external noise, measurement of their travel times, and recovery of their forms; 2) solution of the inverse problem of calculating the parameters of events from measurement data; At present, there exist some successive algorithms, that is, algorithms allowing online determination of wave arrival times with minimization of their measurement errors in (2). The successive approach is oriented to obtaining the "(currently)" fastest but, in the general case, not optimal solution to the problem. It forms a basis for a family of algorithms to detect the times of changes in the properties of signals [2]. Among them is the autoregressive integrated moving average (ARIMA) algorithm [3] used to detect earthquakes [3] and industrial explosions [4]. In recent years, wavelet filtration algorithms have been widely used to solve problems of geophysical monitoring and analyze seismic data [5–7]. In the class of considered algorithms traditionally used a deconvolution, based on Wiener-Hopfa inverse filtration. The last set as the purpose approach seismic impulse to δ a-shaped kind [8]. This list of successive algorithms is far from complete. In addition to the successive algorithms, there are also a posteriori (off-line) algorithms. In contrast to the former, the latter are oriented to obtaining an optimal solution (solution over all accumulated data). In other words, this approach is potentially more accurate than the successive one. However, its algorithmic implementation involves solving discrete optimization problems using cumbersome calculations. Therefore, most existing off-line technologies for solving such problems have several stages (subproblems): for instance, noise filtering with subsequent solution of the problems of detection, estimation, or decision-making. A key shortcoming of step-by-step data processing is as follows: even in the case of optimal solution of subproblems at each stage the resulting solution may not coincide with the optimal one, because, in the general case, the solution found on the basis of conditional extremums must not coincide with the optimal one. In the present paper, another approach, which has been insufficiently studied as applied to geophysical monitoring, is investigated. Within the framework of this approach, a solution to the problem is found in a unified process of discrete optimization without dividing the problem into stages. The following two types of detection are possible: direct estimation of wave arrival times or simultaneous obtaining of arrival time estimates and wave pulse shapes. In this class of algorithms, we will consider those designed for the processing of sequences that change their properties *quasi-periodically* [9,10]. This means that the time interval between two successive pulses is bounded from above and below by given constants.

2.2 Posteriori Algorithms for Determining the Parameters of Wave Forms in Noise

In this section, the a posteriori algorithms for solving the problems of detection and separation of waveforms presented by a quasi-periodic sequence and distorted by Gaussian noise are justified. We consider two variants of waveforms in a quasi-periodic sequence, both identical and different. To solve the problem, the following model of data for analysis is proposed. Let the vector components $X = (x_0, ..., x_{N-1}) \in R^N$ form the sequence

$$x_n = \sum_{m=1}^{M} u_{n-n_m}(m), \; n = 0, ..., N-1, \; (n_1, ..., n_M) \in \Omega_M, \tag{3}$$

$$\Omega = \bigcup_{M_{\min}}^{M_{\max}} \Omega_M,$$

$$\Omega_M = \{(n_1, ..., n_M) \mid 0 \leq n_1 \leq T_{\max} - q; \quad N - T_{\max} \leq n_M \leq N - q; \atop q \leq T_{\min} \leq n_m - n_{m-1} \leq T_{\max}, \quad m = 2, ..., M\}, \tag{4}$$

Assume that $u_j(m) = 0$, if $j \neq 0, ..., q-1$, at each $m = 1, ..., M$, and M_{min} and M_{max} are found from the solution to the systems of inequalities in the definition (4), in which q, T_{min}, T_{max} are natural numbers. Also assume that $U_m = (u_0(m), ..., u_{q-1}(m)), m = 1, ..., M, w = (U_1, ..., U_M)$ and $\eta = (n_1, ..., n_M), m = 1, ..., M$. Let $0 < \|U_m\|^2 < \infty, m = 1, ..., M$. Then, according to the introduced notation, the vector X depends on the pair of sets η and w having the same number of M elements, that is, $X = X(\eta, w)$. Let the random vector $Y = (y_0, ..., y_{N-1})$ be the sum of two independent vectors, $Y = X(\eta, w) + E$, where $E = (e_0, ..., e_{N-1}) \in \Phi_{X,\sigma^2 I}, \sigma^2 < \infty$. Here $\Phi_{X,\sigma^2 I}$ denotes normal distribution with the parameters $(0, \sigma^2 I)$.

With allowance for the above, the problem of detection of quasi-periodic sequences of waveforms is in finding, with the observed vector Y, the set η, according to which the non-observed vector $X(\eta, w)$ was generated. In this model, components of the vectors Y and X correspond to the observed and non-observed signals, and components of the vector E, to noise. The numbers of the vector components are associated with uniform discrete time. Elements of the set $(n_1, ..., n_M)$ correspond to the arrival times of waveforms, and the q-dimensional set $U_m, m = 1, ..., M$, corresponds to a waveform. The values of T_{min} and T_{max} are interpreted as the maximum and minimum intervals between two successive forms. To solve such problems, the principle of maximum likelihood is used. It has been shown by the authors that noise-immune maximum likelihood detection of a given number of unknown wave forms can be simulated by the following discrete extremal problems:

Problem 1. Given: a numerical sequence $Y = (y_0, ..., y_{N-1})$, natural numbers q, M, T_{min}, T_{max}. Required: a set $\eta = (n_1, ..., n_M) \in \Omega_M$ such that

$$F(n_1, ..., n_M) = \sum_{m=1}^{M} \sum_{k=0}^{q-1} y_{n_m+k}^2 \rightarrow \max.$$

In the case that all waveforms are identical, that is, $U_m = U = (u_0, ..., u_{q-1})$ for every m=1,...,M, and their number M is not known, the problem of detection of these forms induces the following extremal problem:

Problem 2. Given: a numerical sequence $Y = (y_0, ..., y_{N-1})$, a vector $U = (u_0, ..., u_{q-1})$, natural numbers q, T_{min} and T_{max}. Required: a set $\eta = (n_1, ..., n_M) \in \Omega_M$ and its dimension such that

$$S(n_1, ..., n_M) = \sum_{m=1}^{M} \sum_{k=0}^{q-1} u_k(u_k - 2y_{n_i+k}) \rightarrow \min \qquad (5)$$

The functions F and S are separable. Therefore, problems 1 and 2 are exactly solved by the same method of dynamic programming, but using different recurrence formulas. Problem 1 is solved in time $O(MN^2)$, and problem 2, in time $O(N^2)$. It's induced by the problem of joint detection and estimation of the recurrent form at an unknown number of recurrences is most difficult. The problem is formulated as follows.

Problem 3. Given: a numerical sequence $Y = (y_0, ..., y_{N-1})$, natural numbers q, T_{min}, T_{max}. Required: a set $(n_1, ..., n_M) \in \Omega_M$ and its dimension such that

$$G(n_1, ..., n_M) = \sum_{m=1}^{M} \sum_{j=1}^{M} \sum_{k=0}^{q-1} y_{n_m+k} \, y_{n_j+k} \rightarrow \max$$

Optimal values of components of the sought-for set $\hat{U} = (\hat{u}_0, ..., \hat{u}_{q-1})$, corresponding to the waveform are found by the formula

$$\hat{u}_k = \frac{1}{\hat{M}} \sum_{m=1}^{\hat{M}} y_{\hat{n}_m+k}, \; k = 0, ..., q - 1$$

where $n_m, m = \overline{1, \hat{M}}$ and $\hat{M}$ are elements of the optimal solution to problem 3.

This NP problem is difficult. Hence, in the general case its exact solution cannot be found in polynomial time (if $P \neq NP$). Therefore, approximate algorithms are of interest. One of such heuristic algorithms is proposed in the present paper. The idea of the algorithm is as follows: First, find a solution to problem 1 at $M = 1$ for the initial part of the sequence Y containing $T_{max} - q + 1$ elements. Then, using the found value of $\hat{n}_1$, find the set $(y_{\hat{n}_1}, ..., y_{\hat{n}_1+q-1})$. With this set, solve problem 2 setting $U = (y_{\hat{n}_1}, ..., y_{\hat{n}_1+q-1})$. Finally, using the found set $(\hat{n}_1, ..., \hat{n}_{\hat{M}})$, calculate estimates of components of the vector $\hat{U}$. Taking into account the above approach, we propose a two-stage (locally optimal) algorithm for finding an estimate. Namely, at the first stage a rough estimate of the pulse shape is made. At the second stage, this estimate is refined in the process of solving the problem of joint estimation of the pulse shapes and detection of the pulse beginning times. The essence of the first stage is in solving the problem of verifying the hypotheses. The second stage is designed for solving the problem of estimation, which is reduced to minimization of the additive functional.

To verify the performance of the proposed algorithm and study its accuracy, some numerical experiments were made, with simulation of various waveforms and the same forms and duration and complicated by Gaussian noise of different levels. Real waveforms from explosions and vibration sources recorded earlier and various signal/noise ratios were specified. The generated set $(n_1, ..., n_M)$ of random numbers was used to form a sequence of components of the vector X. According to the adopted model, the sequence of components of the vector Y was synthesized as the sum of the vector X and the Gaussian vector E with the distribution parameters $(0, \sigma^2 I)$. As an example, Fig.1 presents, in graphic form, the results of simultaneous detection and separation of waveforms by the algorithm for solving problem 2. This figure shows: a) the generated model noisy sequence and the sequence found by the algorithm for solving problem 2, b) the results of numerical estimation of errors in the separation of identical waveforms in the quasi-periodic sequence on the background of noise for a signal/noise ratio of 1.25. The arrival times for all separated pulses in the both sequences are plotted on the X-axis, at the beginning of each pulse. The series of numerical experiments has shown that the mean absolute error in estimation of the wave-form arrival time is 0.047 s. It is by a factor of 3 smaller than for the wavelet filtration algorithm with a threshold detector used to solve the same problem. To verify the quality of the algorithm for waveform estimation, we used a measure of root-mean-square deviation in the form

$$\delta_U(M) = \frac{1}{q} \cdot \sum_{k=0}^{q-1} (u_k - \hat{u}_k)^2,$$

where $u_k, \hat{u}_k, k = 0, ..., q - 1$ are the given and calculated components of the waveform U. The relative root-mean-square error in the waveform estimation for the data in Fig.1 does not exceed 6 %.

2.3 Fractals in a Posteriori Algorithms

In problems 1-3 and algorithms to solve them, the parameters q, T_{min}, T_{max} corresponding to the waveform duration and the upper and lower bounds of the interval between two successive waveforms are input data. However, in practical problems these parameters are often not known in advance. To remove this a priori indefiniteness, we propose an approach for preliminary estimation of the above parameters based on a fractal representation of waveforms. Waveforms are mapped onto a two-dimensional "frequency-time" plane using a two-dimensional Fourier transform of the form

$$F(k_1, k_2) = \frac{1}{\sqrt{N_1 N_2}} \sum_{n_1=0}^{N_1-1} \sum_{n_2=0}^{N_2-1} F[n_1, n_2] \cdot w_{N_1}^{k_1 n_1} w_{N_2}^{k_2 n_2}, w_N = \exp(-j\frac{2\pi}{N}) \quad (6)$$

Projection of the function of two variables obtained according to (6) onto the "frequency-time" plane will be a two-dimensional image in which the levels of

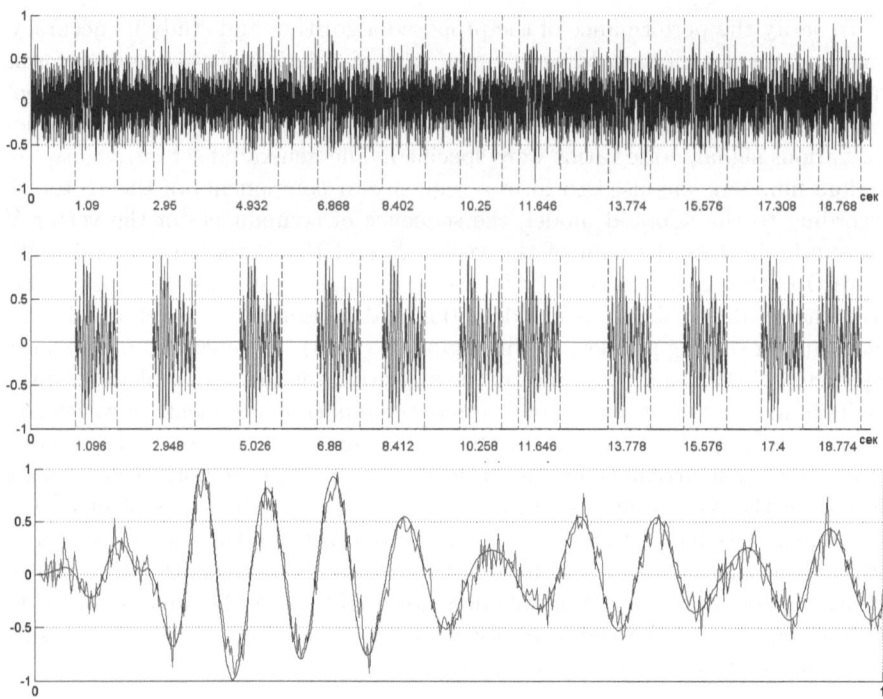

Fig. 1. Signal/noise ratio=1.25, T_{min}=1.3 s, T_{max}=2.2 s, q=1 s; N=20 s, M=11; $\delta_U(\sigma) = 6 * 10^{-2}$

amplitude values will correspond to brightness levels. The thus obtained waveform images serve for preliminary estimation of the wave pulse boundaries. Subsequent corrected calculation is made using the discrete optimization algorithm by solving problem 3. In what follows, the results of a numerical simulation for the fractal approach to separate the waveform boundaries in noise are presented (see Fig.2).

The simulation was made as follows. Real waveforms taken from experiments were specified. The form corresponding to a specific problem was chosen from the set. Then, a frame was formed with different values of the parameters N, M, T_{max}, T_{min} and q according to (3), (4). Noise with a Gaussian distribution with the parameters $(0, \sigma)$ was superimposed on the selected forms. The signal/noise ratio was specified by the level of σ.

Fig.2 gives a qualitative picture of the above. In Fig.2b (top) one can see noisy waveform sequences to be processed. The record contains 8 waveforms cut out from real seismograms from vibrational sources. Fig.2a presents the result of two-dimensional Fourier transform of the record according to (6). Here the starts and ends of wave pulses, including the beginning of a quasi-periodic pulse sequence, are separated well from noise. This improves the performance of the discrete optimization algorithm. Fig.2b (middle) shows waveform sequences with found arrival times, and in Fig.2c (bottom), the dark histogram shows calculation errors of arrival times without the fractal representation (6), and the light histogram, with the fractal representation. The boundaries in the records outline waveform locations, both initial and calculated ones using the fractal approach with the optimization algorithm. It follows from the error plot that the use of the fractal representation of the pulse sequence allows a considerable increase in the accuracy and reliability of determining arrival times by the discrete optimization algorithm. In some cases, the error decreases by an order of magnitude [11].

2.4   Solution of the Inverse Problem

The problem of estimating the parameters $\boldsymbol{\theta}$ in (1) is a part of regression analysis, and its solution are estimates by the least squares method:

$$\boldsymbol{\theta} = \arg\min Q(\boldsymbol{\theta}), \ \ Q(\boldsymbol{\theta}) = \sum_{i=1}^{N} \sigma_i^{-2}(\hat{n}_i - \eta(\gamma_i, \boldsymbol{\theta}))^2 \tag{7}$$

To find a minimum of the functional $Q(\boldsymbol{\theta})$, the Gauss-Newton iterative method or its modifications based on linear approximation of the regression function in the neighborhood of a point $\boldsymbol{\theta}^k$ are used:

$$J(\gamma, \boldsymbol{\theta}^k)\Delta\boldsymbol{\theta}^k + \boldsymbol{\eta}(\gamma, \boldsymbol{\theta}^k) - \boldsymbol{n} + \boldsymbol{\varepsilon} = 0, \tag{8}$$

where

$$J(X, \boldsymbol{\eta}) = \left(\frac{\partial \eta(\boldsymbol{x}_i, \boldsymbol{\theta})}{\partial \theta_1}, ..., \frac{\partial \eta(\boldsymbol{x}_i, \boldsymbol{\theta})}{\partial \theta_m} \right), i = 1, 2, ..., n. \tag{9}$$

To solve equations (7-9), an approach with direct solution of the system (8) at each step of the iterative process by the pseudo- (or generalized) inversion method is used. It is based on singular value decomposition (SVD). It is well-known that the SSVDC procedure in the Linpack library is used to calculate SVD [12]. Paper [13] presents a standard SVD procedure in Fortran-IV used in the present paper. The current MATLAB system versions have a built-in function svd(A) implementing this decomposition for an arbitrary $n \times m$ matrix A. The calculation scheme of the SVD procedure is in decomposing the matrix (9) at each step of the iterative process into the product of three matrices,

$$J(\gamma, \boldsymbol{\theta}^k) = U_k \Sigma_k V_k^T, \tag{10}$$

where U_k is the orthogonal $n \times n$ matrix, V_k is the orthogonal $m \times m$ matrix, and Σ_k is the diagonal $n \times m$ matrix with the structure, where $\Sigma_k = (S_k/0)$, is the diagonal matrix of singular numbers arranged in decreasing order $\rho_i \geq \rho_{i+1}$. The method also includes the so-called singular analysis, which is in excluding zero singular numbers and the corresponding columns of the matrices U and V. In this case the iterative process has the following form:

$$\boldsymbol{\theta}^{k+1} = \boldsymbol{\theta}^k + V_k S_k^{-1} \boldsymbol{d}^k, k = 0, 1, 2, ... \tag{11}$$

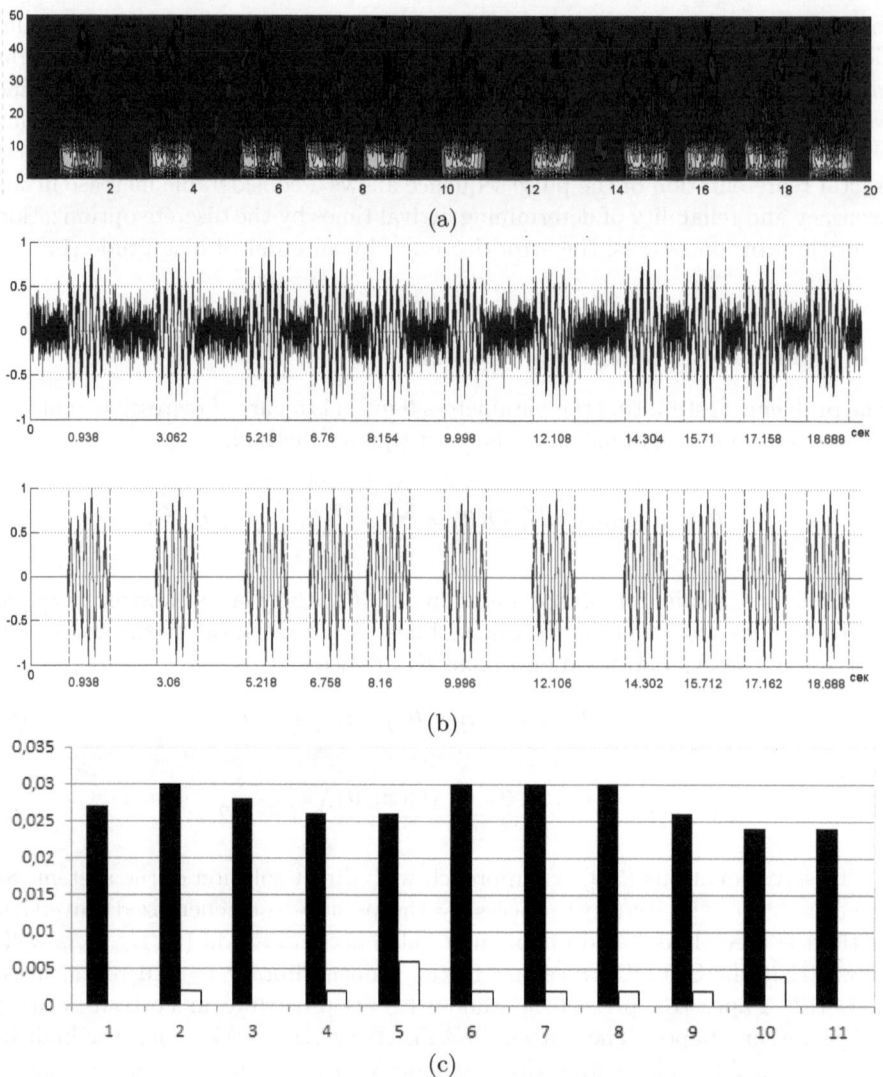

Fig. 2. Signal/noise ratio=3, T_{min}=1.3 s, T_{max}=2.2 s, q=1 s; N=20 s, M=11; $\delta_M(\sigma) = 2*10^{-3}; 7.6*10^{-3}$

where $\boldsymbol{d}^k$ is a vector consisting of the first m components of the vector $U_k^T \boldsymbol{y}(\gamma, \boldsymbol{\theta}^k)$, where $\boldsymbol{y}(\gamma, \boldsymbol{\theta}^k) = (\boldsymbol{n} - \eta(\gamma, \boldsymbol{\theta}))^T$. During this process, not only a covariance matrix of the space of parameters, but also a covariance matrix of the space of data, and a resolution matrix $V_k V_k T$, are obtained. The closeness of the latter matrix to the unit matrix shows the degree of solvability of the problem. The information density matrix is $U_k U_k T$, whose closeness to the unit matrix shows the relative significance of individual observations [14, 15].

3   Conclusions

1. The problem of geophysical monitoring of natural and technogenic events, including environmental monitoring, has been considered. The monitoring has several successive stages, including recording of responses to remote events in the form of seismic waveforms and measurement of their major parameters - arrival times and initial forms. At the final stage, inverse problems of determining the geographical location and time of occurrence of an event are solved.

2. To increase the accuracy of solving the problems of detection and separation of waveforms, a posteriori discrete optimization algorithms have been proposed and analyzed. High accuracy of the proposed algorithms has been proved by numerical experiments. Specifically, it has been shown that the root-mean-square deviation in the estimation of waveforms does not exceed 6 % and relative estimation errors of their arrival times are not worse than 0.1 %.

3. The proposed algorithms were used to solve inverse problems of estimation of the parameters of events, namely, their geographical coordinates and times of occurrence, the velocity characteristic of the medium.

References

1. Nolet, G. (ed.): Seismic Tomography, p. 415. D. Reidel Publishing Company, Dordrecht (1987)
2. Basseville, M., Benveniste, A. (eds.): Detection of Abrupt Changes in Signals and Dynamical Systems, p. 278. Springer-Verlag, Heidelberg (1986)
3. Nikiforov, I.V.: Successive Detection of changes in the properties of time series. Nauka, Moscow (1983)
4. Glinsky, B.M., Khairetdinov, M.S., Omelchenko, O.K., Rodionov, Y.I.: On a new automized technology for seismic source location. Bulletin of the Novosibirsk Computing Center, Series: Mathematical Modeling in Geophysics (8), 35–42 (2003)
5. Khairetdinov, M.S., Avrorov, S.A., Livenets, A.A.: Computing technology in seismic monitoring networks and systems. Bulletin of the Novosibirsk Computing Center, Series: Mathematical Modeling in Geophysics (13), 51–69 (2010)
6. Lubushin, A.A.: Wavelet-aggregated signal and synchronous peaks in problems of geophysical monitoring and earthquake prediction. Fizika Zemli (3), 20–30 (2000)
7. Sagaidachnaya, O.M., Dunaeva, K.A., Sal'nikov, A.S.: Decomposition and analysis of seismic fields on the basis of layers of wavelet-decomposition. Geofisika (5), 9–17 (2010)
8. Hatton, L., Worthington, M.H., Makin, J.: Seismic data processing, 212 p. Blackwell Scientific Publications, London (1986)
9. Kel'manov, A.V., Jeon, B.: A posteriori joint detection and discrimination of pulses in a quasiperiodic pulse train. IEEE Trans. Signal Processing $52(3)$, 1–12 (2004)
10. Gruber, P., Todtli, J.: Estimation of Quasiperiodic Signal parameters by Means of Dynamic Signal Modes. IEEE Trans. Signal Processing $42(3)$, 552–562 (1994)

11. Woskoboynikova, G.M.: Determination of the arrival times of the seismic by the dynamic programming method. In: Proceedings of 9th Korean-Russian International Symposium on "Science and Technology" (KORUS 2005), pp. 734–737. NSTU, Novosibirsk (2005)
12. Kahaner, D., Moler, C., Nash, S.: Numerical Methods and Software, p. 575. Prentice Hall, INC. a Pearson Education Company (1989)
13. Omelchenko, O.K.: Numerical implementation of wave mode of definition of bottom hole coordinates. Bulletin of the Novosibirsk Computing Center, Series: Mathematical Modeling in Geophysics (5), 121–126 (1999)
14. Hudson, D.: Statistics for Physicists, p. 296. Mir., Moscow (1970)
15. Forsythe, J., Malcolm, M., Moler, C.: Machine methods of mathematical calculation, p. 280. Mir, Moscow (1980)

Two-Component Incompressible Fluid Model for Simulating Surface Wave Propagation

Yuri Zakharov[1]($\boxtimes$), Anton Zimin[2], and Vladimir Ragulin[1]

[1] Kemerovo State University, Krasnaya Street, 6, 650043 Kemerovo, Russia
zaxarovyn@rambler.ru, ragulin@gmail.com

[2] Institute of Computational Technologies of the Siberian Branch of the Russian Academy of Sciences, Acad. Lavrentjev Av., 6, 630090 Novosibirsk, Russia
sliii@mail.ru

Abstract. In this paper, the motion model of the two-component incompressible viscous fluid with variable viscosity and density is considered for modeling the process of the surface wave propagation. The model consists of the non-stationary Navier-Stokes equations with variable viscosity and density, the convection-diffusion equation and equations for determining the viscosity and density depending on the concentration of the components. Thus we model the two-component medium, one of the components being more dense and viscous liquid. The results of calculations for two-dimensional and three-dimensional problems are presented.

Keywords: Navier-Stokes equations · Surface wave propagation · Variable viscosity · Variable density · Inhomogeneous fluid · Two-component fluid

1 Introduction

The investigation of the surface waves is a fundamental problem of the hydrodynamics and the environment. Such investigation is necessary for solving a number of applications connected with designing, maintenance and security of ships and coastal structures. The problem of the numerical modeling of such waves (especially those with non-linear character) remains topical despite the fairly large number of studies.

Mesh and mesh-free approaches are used for modeling problems of the wave emergence and propagation on the free surface. A distinctive feature and main advantage of the mesh methods is tracking an interface by nodes of the deformable computational grid. It allows one to describe the free surface as accurate as possible and to account complex boundary conditions (e.g., surface tension). These methods include LINC [1] and ALE [2]. However, using the mesh Lagrangian methods involves considerable difficulty or it is even impossible for the complex flows calculation, where the interface can be destroyed or intersect itself (wave breaking, decay and combining the bubbles, filling reservoirs, etc.).

Mesh-free methods are used as an alternative to the mesh methods. One of them is the mesh-free Lagrangian smoothed-particle hydrodynamics method

© Springer International Publishing Switzerland 2015
N. Danaev et al. (Eds.): CITech 2015, CCIS 549, pp. 201–210, 2015.
DOI: 10.1007/978-3-319-25058-8_20

(SPH) [3-5]. This method has been developed for example in [6, 7]. The particles being not linked by grid lines, they are not limited in their movements in space. Continuous distributions of physical values in the area can be approximated according to these values attributed to the particles. Mesh-free methods allow one to carry out calculations of flows with strong deformations of the computational domain boundaries allowing change of the area connectivity and the boundaries overlap. SPH disadvantage compared with mesh methods is that a large number of particles is necessary to generate a simulation with equivalent resolution. However, the calculation accuracy can be significantly increased by using SPH together with grid-based techniques.

An example of such an association is the method of marker-and-cell (MAC) [8, 9]. MAC method feature is to use a mixed Euler-Lagrange approach. The area being studied is divided by stationary staggered Euler grid into cells. Lagrangian particles-markers mesh being carried by the velocity field is used to determine the position of the free surface and to visualize the flow simultaneously. MAC allows one to calculate complex flows with the uniting and/or dividing fluid volumes, modelling such process with Lagrangian mesh methods being a considerable complexity. This MAC method ability is due to the markers monitoring exactly fluid volume movement, but not that of its surface. The disadvantage is that very large number of particles-markers is needed to be calculated. Lack of the particles reduces the accuracy of determining the contact boundary position and moreover can cause non-physical appearance of a "blank" cells in a volume filled with a liquid (e.g., near a flow stagnation point).

In contrast to MAC, a special phase function was suggested to be used instead of calculating big amount of particles in the method Volume of Fluid (VOF) [10]. Value of this function equals one if the phase is at the point, there being no phase it equals nil. Currently this approach is rather widespread (e.g., [11, 12]). The advantages of VOF method are low computational requirements and the theoretically possible conservatism. Its disadvantage is schematic phase dispersion, which is the consequence of the front smearing of VOF function due to numerical viscosity.

Method Level Set [13] was developed in order to overcome this effect. It also uses a special function of the distance to the free surface. Special function (e.g., Heaviside step function) is used to set the discontinuity of density and viscosity at the interface. The advantage of this method is a good accuracy in determining the geometric shape of the contact boundary. However, the method is poorly applied exactly for applications where the liquid dispersion and fragmentation is physically possible. In addition, since the level function is not explicitly included in the equations of conservation there may be imbalance of mass, momentum, etc. It should be also noted that level functions method cannot be extended in the case of several (more than two) immiscible liquids in contrast to VOF method. Level Set method is also widely used at the present time (e.g., [14, 15]).

It is worth noting that attempts are also made to develop a joint Level Set and VOF method combining the advantages of both approaches [16].

The motion model of the two-component incompressible viscous fluid with variable viscosity and density is considered for modeling the process of the surface wave propagation in this paper. The model consists of the non-stationary Navier-Stokes equations with variable viscosity and density, the convection-diffusion equation and equations for determining the viscosity and density depending on the concentration of the components. Thus we model the two-component medium, one of the components being more dense and viscous liquid.

Previously this model has been used in problems of substance diffusion in the branched channel [17] and cohesive soil erosion [18].

An important step in the model is the calculation of the occurring medium movement. Methods for calculating the stationary and non-stationary flow problems have been considered in [19, 20]. Methods for solving problems with given pressure difference as the boundary conditions have been considered and two-dimensional and three-dimensional calculations have been carried out in [21, 22]. The variable viscosity has been used to accelerate the convergence rate of iterative schemes for solving problems of viscous incompressible flow in [23].

2   Mathematical Model

The motion of medium consisting of two incompressible miscible liquids with densities ρ_1, ρ_2 and viscosities μ_1, μ_2 is considered. We mean a solution $\overline{x} = \overline{x}(t)$ of the Cauchy problem $\frac{d\overline{x}}{dt} = \overline{V}(\overline{x}, t)$, $\overline{x}(0) = \overline{x}_0$ by mixture particle, where $\overline{V}(\overline{x}, t)$ is a velocity vector of the mixture at the point $\overline{x} = (x_1, x_2, x_3)$ and time momentum t. $C(\overline{x}, t)$, μ and ρ denote the volume concentration of one component (more dense and viscous), the dynamic viscosity and the mixture density respectively.

The following dependencies on the components concentration are used for finding the viscosity and density of the mixture:

$$\begin{cases} \mu = C\left(\mu_2 - \mu_1\right) + \mu_1, \\ \rho = C\left(\rho_2 - \rho_1\right) + \rho_1. \end{cases} \tag{1}$$

Mass diffusion occurs between the particles of the mixture according to the law:

$$q_n = -\left(\rho_2 - \rho_1\right) D \frac{\partial C}{\partial \overline{n}}, \tag{2}$$

where D is the diffusion coefficient.

Moving volumes ω_t do not change its value in time due to the mixture incompressibility:

$$\int_{\omega_t} 1 \, dx = const, \tag{3}$$

hence

$$div\left(\overline{V}\right) = 0. \tag{4}$$

The mass balance equation for the fluid volume ω_t:

$$\frac{d}{dt} \int_{\omega_t} \rho \, dx = - \int_{\partial \omega_t} q_n \, d\sigma. \qquad (5)$$

Equality can be obtained from (Eq. 5):

$$\frac{d\rho}{dt} + \rho \, div\left(\overline{V}\right) = (\rho_2 - \rho_1) \, D \Delta C \qquad (6)$$

or taking into account (Eq. 1) and (Eq. 4)

$$\frac{dC}{dt} = D \Delta C. \qquad (7)$$

From the integral momentum equation

$$\frac{d}{dt} \int_{\omega_t} \rho \overline{V} \, dx = \int_{\partial \omega_t} P_n \, d\sigma + \int_{\omega_t} \rho F \, dx \qquad (8)$$

taking into account (Eq. 4), (Eq. 6) and known relation

$$\frac{d}{dt} \int_{\omega_t} \rho F \, dx = \int_{\omega_t} \left[\frac{d}{dt} \left(\rho F \right) + \rho F div\left(\overline{V}\right) \right] dx \qquad (9)$$

the following equation is obtained:

$$\rho \frac{d\overline{V}}{dt} = -\overline{V} \left(\rho_2 - \rho_1 \right) D\Delta C + div P + \rho \overline{F}, \qquad (10)$$

where P is the stress tensor in the mixture, $\overline{F} = (f_1, f_2, f_3)$ is the vector of mass forces.

Then the equations system for the motion of two miscible incompressible fluids mixture taking into account variable viscosity is obtained:

$$
\begin{cases}
\frac{\partial v_i}{\partial t} + \sum\limits_j v_j \frac{\partial v_i}{\partial x_j} = \frac{1}{\rho} \left(-v_i \left(\rho_2 - \rho_1 \right) D\Delta C - \frac{\partial p}{\partial x_i} + \right. \\
\left. + \frac{\partial}{\partial x_i} \left(2\mu \frac{\partial v_i}{\partial x_i} \right) + \sum\limits_{j \neq i} \frac{\partial}{\partial x_j} \left(\mu \left(\frac{\partial v_i}{\partial x_j} + \frac{\partial v_j}{\partial x_i} \right) \right) \right) + f_i, \ i = 1, 2, 3, \\
\sum\limits_{j=1}^{3} \frac{\partial v_j}{\partial x_j} = 0, \\
\mu = C \left(\mu_2 - \mu_1 \right) + \mu_1, \\
\rho = C \left(\rho_2 - \rho_1 \right) + \rho_1.
\end{cases} \qquad (11)
$$

where p is pressure in the mixture.

Thus the model given consists of the convection-diffusion equation for the concentration of the components, relations to determine the density and the viscosity coefficient and hydrodynamic Navier-Stokes equations for incompressible viscous fluid.

A pressure difference and tangential velocity components or the total value of the velocity vector as the boundary conditions at the inlet and outlet is set for motion equations. We use a no-slip condition on the solid wall and boundary conditions of the second kind for the concentration equation. Some initial distribution for the concentration is also given [18].

3   Solution Scheme

To solve the initial boundary problem (Eq. 11) we used the following algorithm. It comprises three stages. The time step for the hydrodynamic part of the equations system (Eq. 11) is done in the first stage, based on the known velocity and concentration distribution (and hence the density and viscosity). The scheme of splitting on physical factors [24] with variable density is used for this purpose:

$$
\begin{aligned}
&\frac{\tilde{V}-V^n}{\Delta t} = -\left(V^n \cdot \nabla\right) V^n + \frac{1}{\rho}\left(-V^n \left(\rho_2 - \rho_1\right) D\Delta C + \mu \Delta V^n + \right. \\
&\left. + \left(\nabla \mu \cdot \nabla\right) V^n + \left(\nabla \mu \cdot J_{V^n}\right)\right) + \overline{F}, \\
&\rho \Delta p^{n+1} - \left(\nabla \rho \cdot \nabla p^{n+1}\right) = \frac{\rho^2 \nabla \tilde{V}}{\Delta t}, \\
&\frac{V^{n+1}-\tilde{V}}{\Delta t} = -\frac{1}{\rho}\nabla p^{n+1}.
\end{aligned}
\tag{12}
$$

The first equation of system (Eq. 12) is considered to describe the transfer of momentum only by convection and diffusion. Thus the intermediate velocity field does not satisfy the continuity equation. However, this field maintaining the

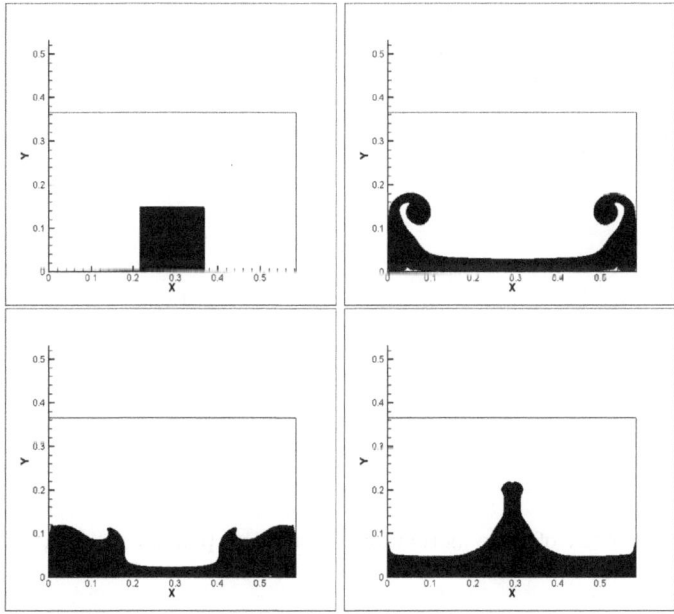

Fig. 1. Picture of wave motion for various time points $t = 0, 0.5, 0.7, 1.0$.

vortex characteristics of the interior points, it has got a physical significance. The following two equations describe the transfer of momentum only by the pressure gradient taking into account the continuity equation.

The time step for the convection-diffusion equation (Eq. 7) is done in the second stage, using the values obtained for the velocity components. We use a predictor corrector scheme with approximation of the convective terms against the flow [25] for this purpose.

The values of the density and viscosity in the space are recalculated according to (Eq. 1) in the third stage. Then the transition to the first stage of the next iteration of the algorithm follows.

It is worth noting that the system of equations (Eq. 11) is solved numerically by the grid method on the staggered grid [26].

Solving an algebraic system of equations obtained as a result of the equation discretization for finding pressure in (Eq. 12) represents one of the most important and dominant moments of the computational procedure in terms of the computing cost. The task can be very complicated because of the operator of this system often being nonselfadjoint and indefinite. The biconjugate gradient stabilized method (BiCGStab) [27] was used to solve this part of the computing process.

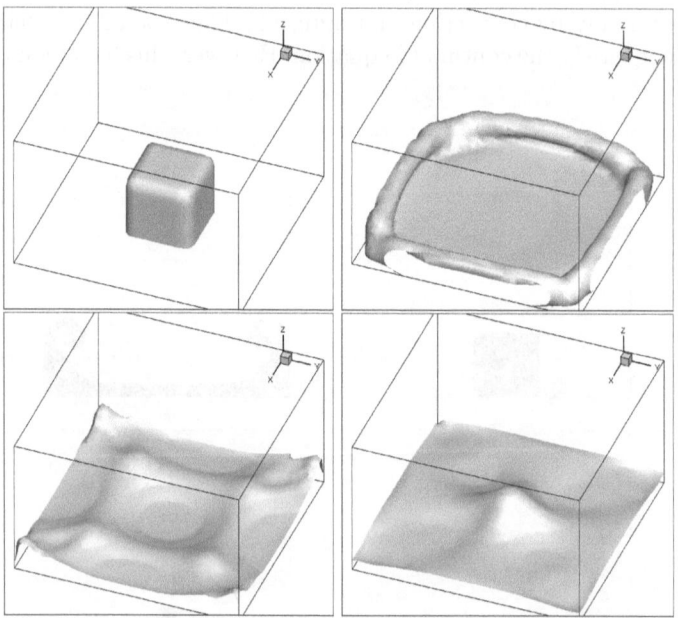

Fig. 2. Picture of wave motion for various time points $t = 0, 0.5, 0.9, 1.3$.

4   Wave Propagation on Surface

The two-component fluid model described by (Eq. 11) is used to simulate the propagation of surface waves. Here one of the components (more dense and viscous) simulates the behavior of the fluid, and another one does that of the gas. We consider the boundary of the two components to take place at $C = 0.1$.

We considered the following problems to test the proposed method. The first one is the collapse of the liquid column. The liquid column is in the middle of the area at the initial time. Then column collapses under the influence of the gravity and movement of the entire medium takes place. The following hydrodynamic parameters were chosen here: $\mu_1 = 10^{-3}$, $\rho_1 = 10$ for liquid and $\mu_2 = 10^{-5}$, $\rho_2 = 1$ for gas. All the borders of area are solid. Fig. 1 and Fig. 2 show the appearance of the wave motion for two- and three-dimensional cases respectively.

The second one is the wave overrunning on the obstacle. Rectangle of the liquid substance is located above the general level in the left side of the area at the initial time. Then the collapse of the rectangle launches a wave in the direction of the obstacle. Here we used the same viscosities and densities as in the first problem. All the borders of area are solid. Fig. 3 and Fig. 4 show the wave overrunning on the obstacle for two- and three-dimensional cases respectively.

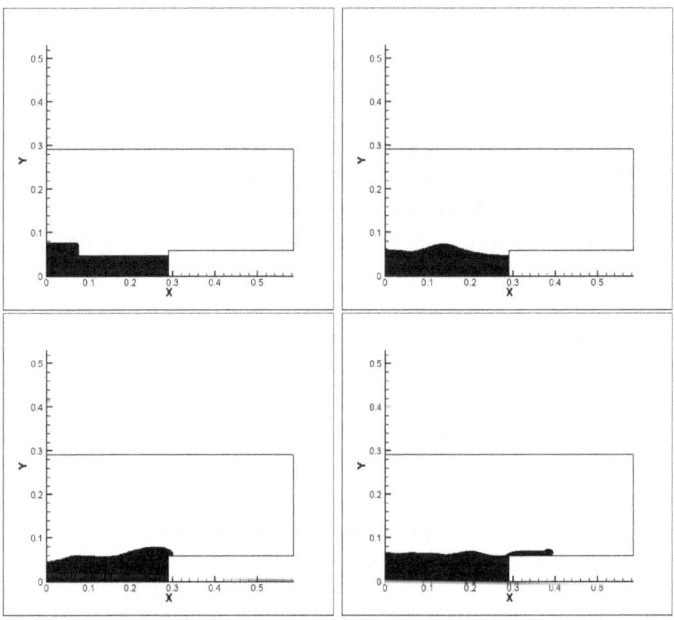

Fig. 3. Picture of wave motion for various time points $t = 0, 0.2, 0.4, 0.7$.

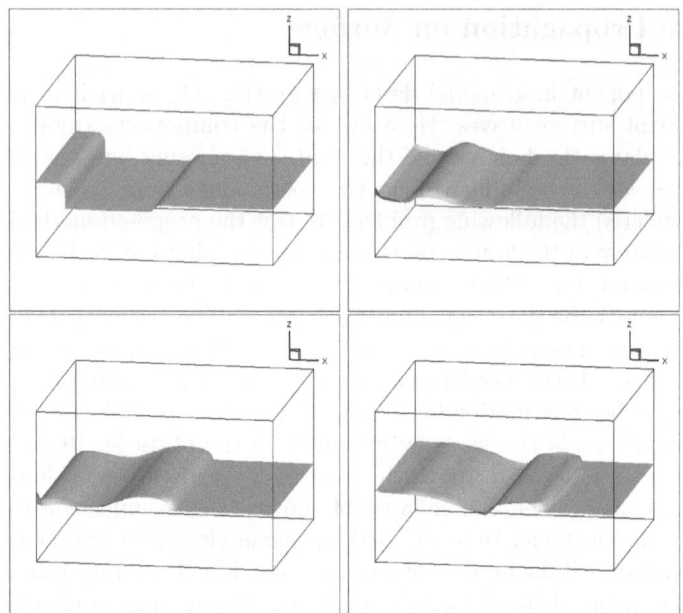

Fig. 4. Picture of wave motion for various time points $t = 0, 0.2, 0.5, 0.8$.

5   Conclusion

The main output stages of the two-component viscous incompressible fluid model were considered and numerical algorithm for solving the resulting model was chosen as well in this work. Calculations for two-dimensional and three-dimensional problems of the wave emergence and propagation on the free surface were carried out.

Acknowledgments. The work was carried out with support of state task of Ministry of Science and Education, project number 1.630.2014/K.

References

1. Butler, T.D.: LINC method extensions. Proceedings of the Second International Conference on Numerical Methods in Fluid Dynamics Lecture Notes in Physics **8**, 435–440 (1971)
2. Hirt, C.W.: An arbitrary lagrangian-eulerian computing method for all speeds. Journal of Computational Physics **14**, 227–253 (1974)
3. Gingold, R.A., Monaghan, J.J.: Smoothed Particle Hydrodynamics: Theory and Application to Non-Spherical Stars. Monthly Notices of the Royal Astronomical Society **181**, 375–389 (1977)
4. Lucy, L.B.: A Numerical Approach to the Testing of Fusion Process. The Astronomical Journal **82**(12), 1013–1024 (1977)

5. Vila, J.P.: On Particle Weighted Methods and Smooth Particle Hydrodynamics. Mathematical Models and Methods in Applied Sciences **9**(2), 161–209 (1999)
6. Potapov, A.P., Rojz, S.I., Petrov, I.B.: Modelirovanie volnovyh processov metodom sglazhennyh chastic (SPH) [Modeling of wave processes by smoothed particle hydrodynamics (SPH)]. Matematicheskoe Modelirovanie **21**(7), 20–28 (2009)
7. Afanas'ev, K.E., Makarchuk, R.S.: Calculation of hydrodynamic loads at solid boundaries of the computation domain by the ISPH method in problems with free boundaries. Russian Journal of Numerical Analysis and Mathematical Modelling **26**(5), 447–464 (2011)
8. Harlow, F.H., Welch, J.E.: Numerical Calculation of Time-dependent Viscous Incompressible Flow of Fluid with Free Surface. Phys. Fluids (American Institute of Physics) **8**(12), 2182–2189 (1965)
9. McKee, S., Tome, M.F., Ferreira, V.G., Cuminato, J.A., Castelo, A., Sousa, F.S., Mangiavacchi, N.: The MAC method. Computers & Fluids **37**, 907–930 (2008)
10. Hirt, C.W.: Volume of fluid (VOF) method for the dynamics of free boundaries. Journal of Computational Physics **39**, 201–226 (1981)
11. Khrabryi, A.I., Zaitsev, D.K., Smirnov, E.M.: Chislennoe modelirovanie techenii so svobodnoi poverkhnost'yu na osnove metoda VOF [Numerical simulation of flows with free surface based on VOF method]. Trudy Krylovskogo gosudarstvennogo nauchnogo tsentra **78**(362), 53–64 (2013)
12. Yakovenko, S.N., Chan, K.S.: Approksimatsiya potoka ob"emnoi fraktsii v techenii dvukh zhidkostei [Approximation of volume fraction stream in flow of two liquids]. Teplofizika i aeromekhanika **15**(2), 181–199 (2008)
13. Osher, S.: Front propagating with curvature-dependent speed: algorithms based on Hamilton-Jacobi formulations. Journal of Computational Physics **79**, 12–49 (1988)
14. Tonkov, L.E.: Chislennoe modelirovanie dinamiki kapli vyazkoi zhidkosti metodom funktsii urovnya [Numerical simulation of the dynamics of viscous liqued drop by level set method]. Vestnik Udmurtskogo Universiteta **3**, 134–140 (2010)
15. Nikitin, K.: Realistichnoe modelirovanie svobodnoi vodnoi poverkhnosti na adaptivnykh setkakh tipa vos'merichnoe derevo [Realistic simulation of the water surface on adaptive grids of octal tree type]. Nauchno-Tekhnicheskii Vestnik SPbGU ITMO **70**(6), 60–64 (2010)
16. Sussman, M., Puckett, E.G.: A Coupled Level Set and Volume of Fluid Method for Computing 3D and Axisymmetric Incompressible Two-Phase Flows. Journal of Computational Physics **162**, 301–337 (2000)
17. Gummel, E.F., Milosevic, H., Ragulin, V.V., Zakharov, YuN, Zimin, A.I.: Motion of viscous inhomogeneous incompressible fluid of variable viscosity. Zbornık radova konferencije MIT **2013**, 267–274 (2014)
18. Zakharov, Y., Zimin, A., Nudner, I., Ragulin, V.: Two-Component Incompressible Fluid Model for Simulating the Cohesive Soil Erosion. Applied Mechanics and Materials **725–726**, 361–368 (2015)
19. Zakharov, Y.N., Ivanov, K.S.: Ob ispolzovanii gradiyentnykh iteratsionnykh metodov pri reshenii nachalno-krayevykh zadach dlya trekhmernoy sistemy uravneniy Navye-Stoksa [Gradient iterative methods for solving initial boundary value problems for three-dimensional Navier-Stokes equations]. Vychislitelnyye tekhnologii **16**(2), 55–69 (2011)
20. Balaganckii, M.Y., Zakharov, Y.N., Shokin, Y.I.: Comparison of two- and three-dimensional steady flows of a homogeneous viscous incompressible fluid. Russian Journal of Numerical Analysis and Mathematical Modelling **24**(1), 1–14 (2009)

21. Milosevic, H., Gaydarov, N.A., Zakharov, Y.N.: Model of incompressible viscous fluid flow driven by pressure difference in a given channel. International Journal of Heat and Mass Transfer **62**, 242–246 (2013)
22. Geidarov, N.A., Zakharov, Y.N., Shokin, Y.I.: Solution of the problem of viscous fluid flow with a given pressure differential. Russian Journal of Numerical Analysis and Mathematical Modelling **26**(1), 39–48 (2011)
23. Janenko, N.N., Shokin, J.I., Zaharov, J.N.: On the nonlinear acceleration of iterative schemes. In: Quatrieme Colloque International sur les Metodes de CalculScientifiqueet Technique, France, p. 20 (1979)
24. Belotserkovskiy, O.M.: Chislennoye modelirovaniye v mekhanike sploshnykh sred [Numerical modeling in continuum mechanics]. Fizmatlit, Moscow (1994)
25. Yanenko, N.N.: Metod drobnykh shagov resheniya mnogomernykh zadach matematicheskoy fiziki [Method of fractional steps for solving multidimensional problems of mathematical physics]. Nauka, Novosibirsk (1967)
26. Patankar, S.: NumericaL Heat Transfer and Fluid Flow. Hemisphere Publishing Corporation (1980)
27. van der Vorst, H.A.: Bi-CGStab: a fast and smoothly converging variant of Bi-CG for the solution of nonsymmetric linear systems. SIAM Journal on Scientific and Statistical Computing **13**, 631–644 (1992)

Author Index